P9-DNA-997

# THE PAPER REBELLION

# THE PAPER REBELLION

*Development and Upheaval in Pulp and Paper Unionism*

Harry Edward Graham

UNIVERSITY OF IOWA PRESS IOWA CITY

*Library of Congress Catalog Card Number: 79-131059*
*University of Iowa Press, Iowa City 52240*

*Printed in the United States of America*
*ISBN 87745-019-6*

# CONTENTS


Preface vii

Acknowledgments ix

Introduction xi

1. Unionism in the Pulp and Paper Industry 1

2. Organization on the West Coast 25

3. Ferment on the West Coast 43

4. Rising Discontent 61

5. Challenge to International Control 71

6. The Detroit Convention: Defeat of the Rank and File Movement for Democratic Action 99

7. The Breakaway on the West Coast 121

8. Schism and Its Implications 145

Bibliography 159

Index 163

# PREFACE

The disease of structural rigidity is no respecter of institutions: Congress, the Pentagon, political parties, the corporation, the trade union, the church, the university. But within most of these institutions are some concurrent tendencies which work in the opposite direction, jolting, wracking the organization, forcing it to respond to new situations.

Unions are particularly vulnerable to the disease of rigidity because the union is by nature a "fighting" organization—that is in its day-to-day life it must husband its resources for frequent life and death battles with a formidable enemy, the employer. Consequently, the union stresses "unity," "loyalty to incumbent officers." Union dissidents are constantly subject to the accusation of "treason," "collaboration with the enemy." When this repressed or neglected dissent finally does surface it may do so in an overly-acrimonious manner, often more so than the real differences between parties would warrant.

Among trade unions some of the common challenges to the status quo take the following forms: (1) Hard fought election contests challenging incumbents (some critics to the contrary, such contests are common in trade unions at all levels); (2) emergence of unions among staff employees within the international; (3) creation of racial pressure blocks within international unions; (4) changes of union affiliation by disgruntled locals or segments of the union; and (5) increasing use by dissidents of the rights granted members under the Landrum-Griffin, Taft-Hartley, and Wagner Acts, often resulting in court interference in union management.

The importance of Mr. Graham's book is that it describes a unique form of dissent by a large group of locals against the two principal international unions in the pulp and paper industry, that of severing ties with the international and continuing as an independent group rather than affiliating with a rival union. Indeed, almost every other disaffiliation has been abetted and usually financed by a rival union.

The situation described by Mr. Graham is unique in another way, the size of the disaffiliation. Twenty thousand workers in two international unions broke away from their affiliated internationals.

Mr. Graham's description is objective. He had access to rich sources both from the dissident unions and from the largest of the international unions. While he has clearly stated the positions of each side, he has perhaps wisely not attempted to evaluate the merits of the action. The contemporaneity and fluidity of the situation more or less dictated this course. In another ten years, it is to be hoped that Mr. Graham will write a postscript to this work, a subjective evaluation of the events.

*School for Workers* *Robert Ozanne*
*The University of Wisconsin*
*1970*

# ACKNOWLEDGMENTS

The author of this book is indebted to far too many people and organizations to permit a complete listing, for a study such as this can only be conducted with the cooperation of the institutions and persons involved. The assistance of the following organizations and individuals is acknowledged with thanks: The Association of Western Pulp and Paper Workers, particularly Secretary-Treasurer Burt D. Wells who graciously donated the papers of the Association to the Wisconsin State Historical Society; the International Brotherhood of Pulp, Sulphite and Paper Mill Workers, its President-Secretary Joseph Tonelli, its Office Manager Francis Tierney, and its former Research Director, John McNiff—they assisted the author in every possible fashion by permitting unrestricted access to the files of the union, as well as giving hours of their time for interviews and discussion, and without their interest and assistance this study could not have been undertaken; the Wisconsin State Historical Society, particularly Manuscript Curator Josephine L. Harper and her staff; the School for Workers of The University of Wisconsin, especially Professor Robert Ozanne whose interest in the project and helpful suggestions were invaluable, Professor George Hagglund who carefully reviewed and criticized earlier drafts of the manuscript, and Mrs. Betty Jane Reis who assisted in assembling the material. It is doubtful the project could have been conducted without Mrs. Reis's participation; the Center for Labor and Management of The University of Iowa, particularly Professor Don Sheriff who read the manuscript and made several helpful suggestions.

My wife, Joyce Ann Graham, typed the first draft of the manuscript. Without her cheerful encouragement and uncomplaining help, this study could not have been completed.

# INTRODUCTION

More than two decades ago, the late Chief Justice Stone commented that unions are

> clothed with power not unlike that of a legislature which is subject to constitutional limitations on its power to deny, restrict, destroy, or discriminate against the rights of those for whom it legislates and which is also under an affirmative constitutional duty to protect those rights.[1]

If Justice Stone's conception of the union as a legislature is correct, it implies that it is reasonable to expect the union to be democratic in nature. Democracy is an ethic which permeates American life. Not only government, but private institutions too, are presumed to be democratic.

The unions proclaim themselves to be democratically run organizations, and the AFL-CIO Ethical Practices Code states:

> Freedom and Democracy are essential attributes of our movement. Labor organizations lacking these attributes are unions in name only. Authoritarian control is contrary to the spirit, the tradition, and the reasons which should always guide and govern our movement.[2]

The comments of Justice Stone and the expression of a democratic standard by the AFL-CIO mirror a conception of the trade union that is apparently held by the general public. While there is little demand that other types of organizations such as the corporation, the professional association, or the fraternal lodge be democratically governed, a separate standard is applied to unions. Thus the exposure of autocratic control in the late 1950s brought forth immediate demands for legislation ensuring the protection of membership rights and civil liberties within the unions.

In the trade union context, democracy is manifest in several areas. First, the membership must have the right and the opportunity to participate in union decisions which affect its welfare. Participation may be through the membership meeting, an elected representative, or by referendum. Regardless of the form, participation requires that the membership be fully and truthfully informed about union affairs, that it be able to dissent from official policy without fear of retaliation, that opposition to and criticism of incumbent officers be tolerated, and that

1. Steele v. Louisville and Nashville Railroad, 323 U.S. 192, 198 (1944).
2. American Federation of Labor-Congress of Industrial Organizations, *Ethical Practices Code.*

members holding minority viewpoints be allowed to challenge the union hierarchy at the polls in an effort to assume power.

Related to the opportunity for challenging the union administration is the necessity for nondiscriminatory treatment of all members. The majority, often identified with the incumbent administration, must not single out certain groups for unfair treatment. Such treatment may be extended to racial, religious, or ethnic groups. It may also extend to political opponents of an incumbent administration.

A final criterion for union democracy would be the provision for aggrieved members to secure speedy and impartial hearing of their complaints. There should be some type of judicial review by a competent, impartial tribunal with no personal or institutional interest in the outcome. The few unions which have instituted public review boards are moving to satisfy this requirement for union democracy.

In ordering their internal affairs, trade unions have tended to emphasize institutional security. No doubt this emphasis on preserving institutional integrity is a reflection of the hostile environment unions have faced. The stress on the integrity of the union is reflected in the opprobrium attached to dual unionism as a crime against the trade union movement. It is often made a violation of the union constitution or bylaws to take any action which might be construed as detrimental to the interests of the union. Thus we find union constitutions containing clauses penalizing members for "attending or participation in any gathering or meeting whatsoever for the purpose of advocating dual unionism, secession (or) schism. . . ."[3] Unions have adopted constitutional clauses of the most general nature which give the administration wide latitude in disciplining members for alleged offenses against the union:

> Charges may be preferred for any disreputable act, conduct unbecoming a union member, for violation of the laws of a local or the International Union, or failure to observe provisions of a labor agreement.[4]

In addition to facing hostile employers and courts for much of their history, unions have faced in this century the danger of communist subversion. Several unions in the 1920s and 1930s were menaced by attempted communist take-over. The expulsion or withdrawal of some CIO affiliates for alleged communist domination about twenty years ago are still relatively recent events. Many union figures of those days continue to play active roles today.

Finally, in stressing institutional security, often at the expense of individual freedom, the labor movement is reflecting, to some degree,

3. International Brotherhood of Electrical Workers, *Constitution*, as amended at the 28th Convention, St. Louis, Mo., September, 1966. Article XXVII, Section 16, p. 81.
4. United Papermakers and Paperworkers, *Constitution*, revised October 1967 by referendum, Article XIX, Section 6, p. 39.

the temper of our times. The communist hunting days of Senator Joseph McCarthy are not long past. Today there is a cry for law and order at the expense of individual liberties. When functioning in such a climate, it is easy to understand how unions would be less than dedicated to individual rights. When the Supreme Court is attacked by U.S. Senators for decisions advancing individual liberties, can we expect the unions to be more dedicated to libertarian traditions?

The prospective members' introduction to the undemocratic aspects of trade unionism may begin before a union secures bargaining rights. Unless there is an independent union on the scene, it is likely there will be no contest between unions in an effort to organize the plant. The merger of the AFL and the CIO removed much competition. Today the choice is frequently between a union and no union, not between two or more unions. The same merger removed an important check on activities of the international which might have found disfavor with local unions. There is no longer, in most cases, a rival international union available to provide a haven for a dissatisfied local union. Today, too, the employer frowns upon local dissent from international policy. He looks to the international to keep the local in line. Thus dissidents are faced with a solid front, employer and international, if they are dissatisfied with their union.

The expansion of collective bargaining territory, from the local plant to area-wide, nation-wide, or industry-wide negotiations has reduced the role of the local union. It no longer plays the central role which it did not long ago. The negotiation of a new agreement may be attended with much formality and use of experts. It may be conducted in a city far from the affected plants and local unions.

In addition to expansion of the territory covered by the collective agreement, the contract itself has become much more complex. Items such as pension plans, job analysis, health and welfare funds, and life insurance have made their appearance in collective agreements. The typical union officer must rely on staff experts or outside consultants for counsel on technical matters of this type. A situation may develop where such experts exercise de facto control over the collective bargaining process and the administration of the agreement. The alleged undue influence of staff personnel was an issue in the election of I. W. Abel to replace David McDonald as president of the United Steelworkers in the mid-1960s.

One reason society expects unions to be conducted democratically is because they reinforce its democratic basis. It may be unrealistic to expect that our government can continue to be democratic if society's institutions become undemocratic. If rank and file union members come to expect, and perhaps desire, that their organizations be administered by technical experts and self-perpetuating oligarchies, it may be only a matter of time until the same principles will come to be accepted in our federal structure. The desire for efficient government may, in the long run, produce less than the best government.

Another argument for more democracy in unions is the experience they provide for rank and file members and local leaders in working within a democratic system. Such training is invaluable if union leaders are to make a transition to public life. And union leadership positions have served to prepare people for participation in public government. If persons with a union background carry over undemocratic attitudes to the public sector, our political system will inevitably suffer.

A democratically-run union is also valuable because it requires the membership to choose among alternatives. These choice-making situations may occur often in the life of the union: for example, when formulating demands for a new agreement; in choosing between different types of insurance and pension plans; in accepting or rejecting management proposals. If union members in their industrial life are required to face up to hard decisions and make wise choices they may be less reluctant to do so when deciding between political candidates. If the rank and file becomes less susceptible to unrealistic demagoguery in union affairs, it may also be better able to resist such demagoguery in public affairs.

In 1964, approximately 20,000 people in the pulp and paper industry on the United States west coast left the two international unions which had held bargaining rights for many years. The two internationals, the United Papermakers and Paperworkers and the International Brotherhood of Pulp, Sulphite and Paper Mill Workers, are both old AFL-CIO affiliates which have organized most of the industry. The collective bargaining arrangement on the coast was considered a model worthy of imitation.[5] There was supposedly a great deal of membership participation in the formulation of negotiating demands, and during the bargaining itself. The two unions were in an apparently impregnable legal position to retain the bargaining rights due to a series of National Labor Relations Board decisions. Wage rates compared favorably to rates being paid for similar jobs in the industry in other sections of the country. In the face of these circumstances, a majority of the coast-wide bargaining unit decided to sever its affiliation with two old, respected AFL-CIO International Unions and found a new union based on free choice and the idea that employees should have the right to choose the leaders of their organization.[6] A new union was formed which gave dissatisfied workers a chance to change their representation. This independent

---

5. Clark Kerr and Roger Randall, *Causes of Industrial Peace Under Collective Bargaining, Crown Zellerbach Corporation and the Pacific Coast Pulp and Paper Industry* (Washington, D.C., 1948).

6. George W. Brooks, Speech delivered before the Third Convention of the Association of Western Pulp and Paper Workers, October 30–November 4, 1967, p. 13.

(non–AFL-CIO) organization was founded emphasizing internal debate, dissent, and criticism in contrast to the more general trade union practice which seems to equate dissent with disloyalty and criticism with treason.

The policy of the new organization places minimal reliance on outside experts and consultants. A do-it-yourself attitude is evident with much membership participation in decision making.[7]

Why did a group of workers feel compelled to fly in the face of the trends of contemporary union development? Why do they emphasize member decision making, criticism of officers, turnover among elected hierarchy, and minimal use of experts? What forces, events, and individuals shaped their thinking? These are the questions we shall attempt to answer.

In order to understand the reasons for the formation of the independent union, it is necessary to follow the development of unionism within the pulp and paper industry, and to trace the history of the Pulp Workers and its sister union, the International Brotherhood of Papermakers. The record of the unions' activities on the West Coast covers the inception of the Uniform Labor Agreement, the area-wide collective bargaining system, and the unrest that prevailed at the local level during the 1930s and 1940s. The remainder of the book traces the continuing manifestations of local unrest on the Pacific Coast in the 1950s. The formation of a reform caucus within the Pulp Workers is detailed along with its fight to achieve its objectives. The failure of the reform groups and the triumph of the International Executive Board at the 1962 convention is examined, along with the reaction of the dissidents to their apparent defeat. The events that led to the breakaway from the international unions in 1964 are detailed, followed by a summary of the litigation and maneuverings that followed establishment of the independent union. A final chapter attempts to evaluate the secession movement to determine its causal factors, and also presents an appraisal of the impact of these events on the structure of collective bargaining.

---

7. *Ibid.*, p. 12.

1

# UNIONISM IN THE PULP AND PAPER INDUSTRY

At the end of 1964, the three principal unions in the paper industry, the United Papermakers and Paperworkers (AFL-CIO), the International Brotherhood of Pulp, Sulphite and Paper Mill Workers (AFL-CIO), and the Association of Western Pulp and Paperworkers (Independent), claimed total membership of 330,048,[1] and represented a sizeable percentage of the industry's total production worker employment of 488,700.[2] But such representation is relatively new, for during the second half of the nineteenth century and the first thirty years of the twentieth, the unions in the pulp and paper industry were small organizations struggling for survival.

The craft of the paper maker is a proud one, calling for the exercise of a high degree of independent judgment. Historically, the highest paid production worker in the mill is the machine tender. The other members of the machine crew, the back tender, the third, fourth, fifth, and sixth hands aspire to become machine tenders.[3] The skill of the paper maker is easily transferable; that is, a machine tender and crew can easily move from one mill to another if they become dissatisfied. Despite the high skill requirement and the ease of mobility, working conditions for the early paper makers were not good. It was common for wages to be cut when the demand for paper declined, and managements attempted to cut prices to keep production at a high percentage of capacity.

One of the prime factors stimulating formation of unions in the industry during the latter part of the nineteenth century was the long hours of work. After direct appeals to employers had been ignored and efforts to achieve regulatory legislation had proved futile, the workers turned to

---

1. U.S. Department of Labor, Bureau of Labor Statistics, *Directory of National and International Labor Unions in the United States, 1965* (Washington, D.C., 1966), pp. 26–28.
2. American Paper Institute, *The Statistics of Paper, 1966 Supplement* (New York, 1966), p. 19.
3. Irving Brotslaw, *Trade Unionism in the Pulp and Paper Industry* (Ph.D. diss., University of Wisconsin, 1964), p. 80.

organization.[4] Before 1907, it was the practice to operate the mills continuously from midnight Sunday to midnight the following Saturday.[5] There were two sets of workers, one on an eleven-hour day tour, the other on a thirteen-hour night tour. The long hours were a result of the economics of the industry, for the cost of machine downtime, plus the cost of startup make breaks in production costly.

Unionism in the pulp and paper industry began in 1884, when a group of machine tenders in Holyoke, Massachusetts, formed Eagle Lodge as a social club for mutual protection and benefit.[6] Eagle Lodge eventually became Eagle Lodge No. 1 of the International Brotherhood of Papermakers.[7] Membership was restricted to the aristocrats of the trade, the machine tenders, showing a narrow, craft-conscious outlook. Other paper workers in the Northeast followed the example of the founders of Eagle Lodge, and established other social clubs.[8] These groups had no particular urge to change industry labor practices, though from time to time they attempted concerted action on vexing problems.

In order to improve conditions in their industry, the Holyoke machine tenders petitioned the Massachusetts legislature in 1887 and 1891 for reform legislation, but they were unsuccessful.[9] Failing in politics, the workers decided organization was the road to change. Eagle Lodge placed an assessment on each member to finance an organizer to spread the word to other mill towns. Organization was especially successful in the Hudson and Black River Valleys of New York and in Wisconsin.[10] In 1893, the American Federation of Labor granted a charter to the United Brotherhood of Papermakers. The jurisdiction of the new union, however, extended only to machine tenders and beater engineers.[11] After some initial organizing success, the UBPM fell upon hard times, and by 1897 there were only three locals.[12] The remnants of the organization then applied to the AFL for a broader charter to widen jurisdiction to all branches of the paper-making trade. This the AFL granted. The machine tenders, however, became dissatisfied, and in 1898 formed the International Paper Machine Tenders Union. Membership problems beset the IPMTU and they soon admitted back tenders. However, back tenders were second-class citizens, relegated to separate lodges, with no repre-

4. James A. Gross, "The Making and Shaping of Unionism in the Pulp and Paper Industry," *Labor History* (Spring 1964), p. 189.
5. *Ibid.*
6. Gross, p. 184.
7. Brotslaw, p. 84.
8. Gross, p. 184.
9. *Ibid.*
10. *Ibid.*, p. 185.
11. Beater engineers supervise the utilization of the pulp used in papermaking.
12. Brotslaw, p. 85.

sentation on the executive board of their union.[13] In 1902, the United Brotherhood and the Paper Machine Tenders merged to form the International Brotherhood of Papermakers.[14]

In the summer of 1901, active organization began among the workers at Fort Edward, New York, a site destined to become the future home of the International Brotherhood of Pulp, Sulphite and Paper Mill Workers. The men at Fort Edward were granted a charter by the AFL as Fort Edward Labour's Protective Union No. 9259, a federal labor union.[15]

In that summer, an itinerant mattress maker named George Johnson came to town. Legend has it that Johnson was forced to keep on the move because he had been blacklisted for union activity. He carried no papers from the American Federation of Labor or any other organization. "The credential that George Johnson carried was the spirit of trade unionism that burned in the heart of this humble workingman."[16] He began to speak out for organization, and then called a meeting to which about fifty workers came. John H. Malin was induced to sign the roll, others followed—enough to secure the charter from the AFL.

After the federal labor union was formed, the members realized that organization of other mills in the vicinity was required. They selected John Malin to visit Hudson Falls and Palmer, New York. He also went to Northampton, Massachusetts, and from there to Bellows Falls and Wilder, Vermont, Franklin and Berlin, New Hampshire, and Livermore Falls and Rumford, Maine.

In April 1902, a convention of pulp workers in the federal labor unions was held in Bellows Falls, Vermont. That was the first convention of the pulp and sulphite workers. The delegates voted to ask for a two and one-half cent per hour wage increase, a sixty-five hour work week for tour workers and a fifty-nine hour work week for day workers—tour men were working seventy-two hours a week, and day workers sixty hours a week.[17]

During this time a man by the name of James F. Fitzgerald from Fort Edward, New York, was active in organizing federal labor unions.[18] About a year later almost 5,000 pulp and paper mill workers had been organized into federal labor unions in New York, Massachusetts, Vermont, New Hampshire, and Maine.[19] By 1903, the pulp worker organizations were

13. Gross, p. 186.
14. The IBPM survived until 1957 when it merged with the CIO Paperworkers to form the United Papermakers and Paperworkers, AFL-CIO.
15. International Brotherhood of Pulp, Sulphite and Paper Mill Workers, *Pulp, Sulphite and Paper Mill Workers Journal*, July-August 1941, p. 2.
16. IBPSPMW, *Proceedings of the Seventeenth Convention*, March 16–19, 1937, p. 20.
17. The description of the formation of the IBPSPMW in those early years comes from *Proceedings of the Seventeenth Convention*, pp. 19–20.
18. IBPSPMW, *Journal*, January 1931, p. 1.
19. IBPSPMW, *Journal*, July-August 1941, p. 2.

large enough to be granted an AFL charter, but President Gompers and the Executive Council of the AFL had doubts about the ability of the pulp locals to govern themselves, since they were made up of unskilled and semiskilled workers.[20] Thus, the pulp worker locals were admitted to the Paper Makers Union, and James F. Fitzgerald became the representative of the pulp, sulphite and paper mill workers on the Paper Makers Executive Board.[21] At that time the name of the organization was changed to the International Brotherhood of Paper Makers, Pulp, Sulphite and Paper Mill Workers.[22] The pulp workers, however, were accorded less than full acceptance by the Paper Makers. They were forced to maintain separate locals and hold meetings separate from those of the skilled groups. In 1905, the union convention passed a resolution barring all but qualified machine tenders from the presidency of the union.[23] Passage of this resolution did not contribute to harmony between the two wings of the union.

In addition to the split between the skilled and unskilled workers, a conflict developed between Fitzgerald, and J. T. Carey, president of the union. Evidently, it was a case of both men wanting to be the highest authority in the organization.[24] The situation became so bad that the Berlin, New Hampshire, Local No. 23 sent letters to all pulp worker locals asking their opinion about seceding from the union. The majority of locals answered affirmatively, and Local No. 23 called a convention at Burlington, Vermont, January 12, 1906.[25] James Fitzgerald was elected convention chairman. The Executive Board of the Paper Makers attended the convention, and President Carey advised against secession. However, he stated that if the pulp workers were determined to become independent, they should remain in the union until the convention scheduled for that May. Carey said he would not oppose withdrawal of the pulp workers in May, and would work for harmonious relations between the two groups.[26] Carey's position was unacceptable to Fitzgerald, and a resolution establishing a separate organization was passed unanimously. Fitzgerald was elected the first president-secretary of the new union,[27] and headquarters were established at Fort Edward, New York, home of Fitzgerald and Treasurer John H. Malin.

Adjournment of the convention saw the start of open war between the Paper Makers and the Pulp, Sulphite Union. The conflict almost

---

20. *Ibid.*
21. IBPSPMW, *Journal,* January 1931, p. 1.
22. *Ibid.*
23. Brotslaw, p. 87.
24. IBPSPMW, *Journal,* January 1931, p. 1.
25. *Ibid.*
26. *Ibid.*, p. 2.
27. *Ibid.*

destroyed both unions.[28] Fitzgerald and Carey spent their time going from place to place, airing their grievances, calling each other names, and spreading dissension and hatred among the workers.[29] "In some cases, the pulp and sulphite workers scabbed on papermakers; and when the opportunity presented itself, the papermakers returned the compliment."[30] In 1906, the paper makers stayed on the job during a pulp workers strike against the West Virginia Pulp and Paper Company at Mechanicsville, New York.[31] They broke the strike and the pulp local at Mechanicsville, but destroyed their own local in the process. In 1908, the pulp workers returned the compliment to the paper makers in a strike against the International Paper Company which had decided to cut wages. The paper makers chose to fight, but the pulp workers remained on the job and helped the company break the strike. By 1909, both unions were almost out of existence.[32]

At the end of 1908, James Fitzgerald resigned as president of the pulp workers, and John Malin succeeded him. With Fitzgerald gone, the door was open for peace between the two groups. In June 1909, a "peace treaty" was concluded between the unions, and remained in force until the merger of the Paper Makers with the CIO Paperworkers in 1957. The treaty contained ten articles, the first of which defined the jurisdiction of each union. The Paper Makers were given jurisdiction over all machine room help and beater engineers in news, bag, and hanging mills. In other mills, the IBPM had the same jobs, plus beatermen, finishers, calender and rotary men, and their helpers. The IBPSPMW had "all other pulp and paper mill help who are not connected with any other international union."[33] A joint conference board of three executive officers from each organization was established to hold regular meetings. The unions also agreed to "do everything within their power to further the interests of the other organization."[34] On July 2, 1909, the American Federation of Labor issued a charter of affiliation to the International Brotherhood of Pulp, Sulphite and Paper Mill Workers. The war had ended.

The next few years were a time of consolidation and rebuilding. The major event of 1910 was another strike at the International Paper Company. After its victory in the struggle of 1908, the company had introduced an elaborate spy system and the yellow-dog contract. In the

---

28. IBPSPMW, *Journal*, April 1929, p. 2.
29. IBPSPMW, *Journal*, January 1931, p. 2.
30. IBPSPMW, *Journal*, April 1929, p. 3.
31. IBPSPMW, *Journal*, January 1931, p. 3.
32. IBPSPMW, *Journal*, April 1929, p. 3.
33. *Ibid.*
34. *Ibid.* The full text of the peace treaty will be found on page 3 of the April 1929 issue of the IBPSPMW *Journal*.

spring of 1910, the unions called a strike and after twelve weeks of united front were successful.[35]

The history of the union is bound up with that of the firms which today occupy an important place in the industry. In the summer of 1915, a strike began at the mills of the St. Regis Paper Company in New York. This strike began at the Deferiet mill and spread to plants in Norwood, Norfolk, Raymondville, and Carthage, New York, and to Donnaconna, Quebec. For two years the strike continued, with the workers living and eating in a tent city, and holding meetings under a circus big top.[36] On February 28, 1917, John P. Burke signed his first agreement as International president-secretary, ending the strike.

In the early 1920s the manufacturers, aided by the postwar depression, attempted to install the open shop. The struggle was centered in the mills and plants of the largest firm, the International Paper Company. The first pulp sulphite local, Fort Edward Local No. 1, was organized at an International Paper Company mill. From that date until April 30, 1921, employees of IP composed the largest group of members in the unions.[37] From 1911 to 1921, the pulp workers had twenty local unions in the mills of IP, all covered by a blanket union shop agreement.[38] By the spring of 1926, the pulp locals in the mills of the International Paper Company had been destroyed, the union's survival made a matter for doubt, and from 1921 to 1937, all mills of the International Paper Company were operated on an open-shop basis.[39]

In January 1921, the delegates to a joint wage conference sponsored by the Pulp Sulphite union and the Paper Makers unanimously voted in favor of asking for a wage increase at the expiration of the agreement then in force.[40] In the spring of 1921, meetings were held between the unions and the International Paper Company and its subsidiary, the Continental Bag Company, and a coalition of nine other firms, led by the St. Regis Paper Company. The spokesman for the latter group was Fred Carlisle, president of St. Regis, and the coalition was known as the "Carlisle group." At the commencement of bargaining, the Carlisle group and IP were represented by a single committee, but a rupture developed in management ranks when the Carlisle group offered to submit the issues in dispute to arbitration and IP refused.[41]

The unions demanded a 10 per cent increase in wages. The manu-

35. IBPSPMW, *Journal*, January 1931, p. 3.
36. IBPSPMW, *Journal*, January-February 1937, pp. 3–4.
37. IBPSPMW, *Journal*, May-June 1937, p. 1.
38. *Ibid.*
39. *Ibid.*
40. The account of the negotiations and strike at International Paper Co. is based mainly upon IBPSPMW, *Proceedings of the Tenth Convention*, October 3–5, 1922, pp. 19–27; and *Proceedings of the Eleventh Convention*, October 7–9, 1924, pp. 17–29.
41. IBPSPMW, *Proceedings of the Tenth Convention*, p. 22.

facturers countered with a proposal to lengthen the work day one hour, to nine hours a day; reduction of hourly rates by 30 per cent; and elimination of premium pay for overtime work. A vote on the manufacturers' proposals resulted in rejection by a vote of 6,388 to 5.[42] There followed "the greatest strike that ever took place in the paper industry in this country."[43] The center of the strike was the IP mill at Palmer, New York. Every day (except Sundays) for the five-year duration of the strike, the workers held meetings.[44] John Burke remarked, "I doubt if any other body of organized workers anywhere in the United States can match this record of being on strike for nearly five years and holding daily meeting during all that period."[45] On July 5, 1921, IP resumed operations on an open-shop basis, using strikebreakers. Though IP commanded much greater resources than the unions, it did not have things all its own way. President Burke was able to report to the 1924 convention that IP was not producing more than 60 per cent of its prestrike tonnage.[46] This was after the strike had been on for three years. The experience of IP in the early days of the strike showed the power of the unions. Production at IP mills ranged from 8 to 18 per cent.[47] Gradually, however, production rose to the 60 per cent noted by Pulp Sulphite President Burke. To counter the strike, the company resorted to the courts. It secured injunctions in eight towns of New Hampshire, Maine, Vermont, and New York, but the injunctions did not run the machines.

In 1920, after meeting all expenses, including dividends to stockholders, IP had a net profit of eleven million dollars. In the first quarter of 1921, the company raised the price of paper $15 a ton, to $130 per ton. Then came the strike. The company books for 1921 showed an item of $12,500,000 for current indebtedness, "which means the cost of the strike."[48]

In order to operate their mills, the company resorted to strikebreakers. The firm scoured the country in a search for help, and the situation became so bad it imported workers from Canada to operate the plants in Maine and New Hampshire. For this IP was fined $5,000 for violation of the immigration laws.[49] President J. T. Carey of the Paper Makers summed up the union case:

> To the union men employed: This fight is your fight, and let us use the slogan so well understood by the Americans during the war. Let us give until it hurts. Because it is giving for the cause of freedom

42. *Ibid.*, p. 21.
43. *Ibid.*, p. 22.
44. IBPSPMW, *Journal*, January-February 1937, p. 4.
45. *Ibid.*
46. IBPSPMW, *Proceedings of the Eleventh Convention*, p. 23.
47. International Brotherhood of Paper Makers, *Paper Makers Journal*, Albany, N.Y., July-August 1921, p 8.
48. IBPM, *Journal*, January 1922, p. 10.
49. IBPM, *Journal*, July 1922, p. 18.

> and democracy and the fight carried on by the employees of the International Paper Company is for freedom and democracy if there ever was an industrial strike for that purpose.[50]

The strike at IP dragged on until March 20, 1926. But by 1922, the skilled trades, represented by the Machinists, Carpenters, Electrical Workers and Firemen had gone back to work, and destroyed whatever slim chance the unions had of winning the strike.[51] The IBPSPMW lost twenty-two locals in the struggle.[52] In fiscal 1920, the union received per capita tax on over 15,000 members, a total of $102,137. In fiscal 1926, tax was paid on about 6,000 members, or $43,497.[53] Strikes, and a transfer of members to the Machinists, Firemen and Electrical Workers unions were responsible for the loss of membership.[54] The union expended $232,638.68 to support the strikers in the mills of International Paper, a large sum for a union in those times.

The situation in the Carlisle mills[55] was somewhat different from that at IP. The firms of the Carlisle group and the unions agreed to submit their "honest differences"[56] to arbitration. The companies and the unions[57] met in June 1921, and established an arbitration board of seven members, three from management, three from labor, and Mr. Frank Irvine, the neutral.[58] The agreement provided that outside unskilled labor would receive a minimum of 40 cents per hour, inside classified labor would take a cut of 16⅔ per cent, and all labor receiving 60 cents per hour or more would be cut 10 per cent.[59] The arbitrators felt that the depression had so reduced the demand for paper that a wage reduction was called for to keep the machines running. The reasoning of the arbitration board was stated as follows:

> It is apparent . . . that there must be a substantial reduction in the rate of wages; if the mills are to continue to operate under present condition, and . . . that such a reduction may now be made without

50. *Ibid.*, p. 19.
51. IBPM, *Journal*, May 1922, p. 16.
52. IBPSPMW, *Proceedings of the Twelfth Convention*, October 5–8, 1926, p. 37.
53. *Ibid.*
54. *Ibid.*
55. The Carlisle group of firms included, by June 1921, eleven firms: Abitibi Pulp and Paper Co.; Spanish River Pulp and Paper Co.; Union Bag and Paper Co.; St. Maurice Paper Co., Ltd.; Minnesota and Ontario Power and Paper Co.; Fort Francis Paper Co.; Hanna Paper Corp.; Cliff Paper Co.; Tidewater Paper Mills Co.; St. Regis Paper Co.; Pettebone-Cataract Paper Co.
56. IBPM, *Journal*, July-August 1921, p. 8.
57. The unions involved were: International Brotherhood of Paper Makers; International Brotherhood of Pulp, Sulphite and Paper Mill Workers; United Brotherhood of Carpenters and Joiners of America; International Brotherhood of Electrical Workers; International Association of Machinists, and the International Brotherhood of Stationary Firemen and Oilers.
58. Board members were: for management, Floyd D. Carlisle, E. B. Murray, and Col. C. H. L. Jones; for the unions, J. T. Carey, John P. Burke, and J. T. Foster.
59. IBPM, *Journal*, July-August 1921, p 9.

> very substantially lowering the compensation of the employees, if such compensation be measured in the purchasing power of money, rather than in dollars and cents.
>
> It is also evident that the welfare of the workers depends upon the aggregate compensation paid for the considerable period rather than upon the rates of wages per hour or per day. A lower rate with steady employment affords greater compensation than a higher rate on part time or with long periods of idleness.[60]

The arbitration agreement provided for a rehearing clause, which came to haunt the unions.

On January 4, 1922, on request from the employers, the arbitration board met in New York City to consider another reduction in wages. The manufacturers argued that they could not pay the scale agreed to the previous year because they had been forced to reduce the price of paper.[61] Wages were cut eight cents for all employees receiving less than 54 cents per hour.[62] The vote on the question was four to three, with the neutral, Frank Irvine voting with management.[63]

In April 1922, the manufacturers again proposed a cut in wages. This time they wanted a 10 per cent reduction for all employees, straight time for all overtime, and "locality rates" for unskilled help.[64] Union resistance to this proposal was so strong that the Carlisle companies changed their offer. The proposal was that all employees receiving 54 cents per hour or over retain their present rates, all overtime rates for outside unskilled help to be abolished, and all rates for unskilled help to be set according to locality rates.[65] The skilled trades, Paper Makers, Carpenters, Machinists, Electrical Workers and Firemen accepted the proposal but the Pulp, Sulphite Workers rejected it. The craft unions then signed separate contracts, leaving the pulp workers to fight alone.[66] President Burke of the IBPSPMW commented:

> I believe that no industry can long remain half union and half non-union, and that unless the organizations in the paper industry are far-seeing enough to refuse to allow the manufacturers to pit one organization against the other, that it will be but a matter of time when we shall have the open shop in all of these mills.[67]

## *Expansion into Canada*

Faced with the unwavering hostility of the International Paper Company, with finances depleted and membership ranks reduced, the unions

60. *Ibid.*, p. 10.
61. IBPM, *Journal*, January 1922, p. 4.
62. *Ibid.*
63. IBPM, *Journal*, May 1922, p. 18.
64. *Ibid.*, p. 16.
65. *Ibid.*
66. *Ibid.*, pp. 16–17.
67. IBPSPMW, *Proceedings of the Tenth Convention*, p. 30.

were forced to look to Canada for potential members. Both the Paper Makers and the Pulp, Suphite Workers concentrated their efforts north of the border in the second half of the 1920s. In May 1925, the IBPM and the Pulp, Sulphite Workers opened a joint office in Three Rivers, Quebec. A full-time organizer was on the payroll from July 1, 1925, to February 27, 1926.[68] More than $2,000 was spent in the organizing campaign, which did not fulfill the expectations of the unions.[69] At the 1926 Convention, President Burke of the IBPSPMW recommended continuation of the Canadian office, and expansion of the organizing drive. Moderate success was achieved. In 1927, the mill of the Thunder Bay Paper Company at Port Arthur, Ontario, was organized and a union shop agreement was signed. This mill was purchased in 1928 by Abitibi Power and Paper Company, which continued the agreement. Also in 1927, the Abitibi mill at Pine Falls, Manitoba, was organized. The next year saw somewhat more progress in Canada, and the start of rebuilding the locals in the United States. In January, a local was established at the Hanesville, New York, mill of the St. Regis Paper Company. In May, the Anglo-Canadian Paper Company mill at Quebec City was organized, and a local at the Herrings, New York, plant of St. Regis was reestablished. The St. Regis mill at Oswego, New York, was also organized. Other Canadian organization efforts in 1928 resulted in establishment of a local at Kapuskasing, Ontario, and Beaupre, Quebec. A local was established at Corner Brook, Newfoundland, but the mill was purchased by International Paper and the union did not prosper.[70]

The gains were somewhat offset by losses sustained in the shut-down of mills in Cheboygan, Michigan, and Chateauguay, New York, where locals disbanded, and in the partial operation of the mills of the Spanish River Pulp and Paper Company, then a part of Abitibi Power and Paper Company.[71]

The outstanding improvement in the situation of the IBPSPMW came in the area of finances. With the heavy expenditures for strike support behind it, the union was able to expand its funds. In the period between the twelfth and thirteenth conventions the union treasury increased by $35,000 to $100,558.[72] There were no strikes between conventions.

The moderate improvement in union finances and the Canadian organization came in the face of declining membership. During the 1920s the IBPM and the Pulp Workers had been losing strength. By 1929, total U.S. membership of the two unions was not much over 9,000, or about 7 per cent of total industry employment.[73]

68. *Ibid.*, p. 37.
69. *Ibid.*
70. IBPSPMW, *Proceedings of the Thirteenth Convention*, March 5–7, 1929, p. 28.
71. *Ibid.*
72. *Ibid.*, p. 29.
73. Brotslaw, p. 125.

The situation in the paper industry was paralleled in other sectors. Industrial unions in particular suffered greatly. Irving Bernstein has commented:

> A significant feature of labor's decline in the twenties is that it struck especially hard at organizations that were predominantly industrial in structure. This was true of the coal miners, of Mine Mill, of the Textile Workers, of the ILGWU, and the Brewery Workers.[74]

On the eve of the Depression, President Burke complained to the convention:

> During the past few years [the] trend toward trustification in this industry has been accelerated. Mergers are now the order of the day. The present trend of the newsprint industry at least would indicate that in a very few years a very few companies will own most of the mills. It is most unfortunate that questions of jurisdiction arise to prevent us from organizing these mills in the only practical manner: that is, along industrial lines.[75]

## *The Depression and its Aftermath*

The depression which began in 1929 had a tremendous impact on the unions in the paper industry. Paper production in the United States reached 11.1 million tons in 1929. In 1932, it had fallen to 7.9 million tons.[76] Consumption per capita dropped from 220.3 pounds in 1929 to 156.9 pounds in 1932.[77] The decline in production, coupled with the increase in funded debt most companies had incurred in the 1920s placed many firms in a precarious financial position.

In 1931, newsprint prices dropped to $37.00 a ton.[78] And in 1932, imports (mainly from Canada) were over 630 thousand tons below their 1929 peak.[79] Per capita consumption dropped almost 20 pounds from 1929 to 1933.[80] To alleviate the situation, the Canadian firms formed the Newsprint Institute of Canada to stabilize prices and spread production. The tactic failed, mainly because the International Paper Company sold newsprint to the Hearst papers at prices below those set by the Institute.[81]

President Matthew Burns of the IBPM made a gloomy report to his convention in March 1931:

> As far as any plan that we have in mind is concerned, I want to say that there has been much said during the past year about the five-day

74. Irving Bernstein, *A History of the American Worker, 1920 to 1933: The Lean Years* (Boston, 1960), p. 86.
75. IBPSPMW, *Proceedings of the Thirteenth Convention*, p. 31.
76. *The Statistics of Paper, 1966 Supplement*, p. 24.
77. *Ibid.*
78. Brotslaw, p. 127.
79. *The Statistics of Paper, 1966 Supplement*, p. 26.
80. *Ibid.*
81. Brotslaw, p. 127.

> working week. During the course of the past two or three months we have discovered, by counselling with many dominating employers, that the five-day week will not bring relief. We find that the greater part of the industry is operating on a three and a half day basis. Those that are operating three to four days would be glad to get five days; they are in favor of it, and those operating six days are inclined to feel that that is their good fortune, and why should they be expected to turn their pay over to somebody else.[82]

Most of the union plants in Canada shut down. As of March 1931, Thunder Bay and Fort William had been down one month; International Falls and Fort Francis thirteen months. The plants at the Soo were down, Espanola was down, most of the plants around Ottawa were down, and Port Alfred was down.[83] By the middle of 1932 most of the large newsprint companies in Canada, except International Paper and the Powell River Company in British Columbia were in bankruptcy or receivership.[84]

In addition to the poor situation in Canada, there was the perennial threat of the South, where wages had fallen to 22 cents per hour.[85] The Advance Paper Company showed the unions their cost of production was $30 a ton in the South, against $60 per ton in the northern mills.[86] The union agreed to a 15 per cent wage reduction; the company struggled on for a while, then shut down and moved to the South.[87]

President Burns report to his convention typifies the despair felt by all:

> It has been mighty discouraging for your officers to see these things slipping away and not having the power to stop them. The employers were helpless. The only hope they had was speeding up and shutting down more machines. They didn't reduce costs by the speeding process; they have shut down the other plants and thereby they have increased their costs.
>
> I have been identified with the paper makers organization for the last 25 years. I have always had in front of me the picture of an organized paper industry. The story I have told you this morning is a rather sad story, after all of the effort that has been put into the movement. We find that we are in a precarious position due to an unsound industry and due to selfishness on the part of the employer and the workers.[88]

The effect of the depression was felt by the work force of the industry, both organized and unorganized.

> A considerable number of companies cut wages several times while some did not cut at all. At the bottom of the depression, two-fifths of

82. IBPM, *Proceedings of the Thirteenth Convention,* March 3–6, 1931, p. 1.
83. *Ibid.*, p. 2.
84. Brotslaw, p. 128.
85. IBPM, *Proceedings of the Thirteenth Convention,* p. 7.
86. *Ibid.*
87. *Ibid.*, p. 8.
88. *Ibid.*

> the companies had cut wages by over 20 per cent, another two-fifths had cut from 10 to 20 per cent, and one-fifth either had cut less than 10 per cent or had not made any reduction at all. The variation in the magnitude of cuts was such that at the depression low of March, 1933, common labor rates for white labor in Northern mills varied from a high of 52 cents an hour in a New York Company which had made only one reduction, to a low of 18 cents an hour in a Michigan Company which made three reductions. However, there was a strong modal concentration of mills which paid between 30 and 40 cents an hour for common labor at the bottom of the depression.[89]

The leadership of the unions were reconciled to acceptance of wage cuts in the face of the poor situation in the industry. In most union mills the wage reductions were only about half as large as those of nonunion mills.[90] In 1931, the IBPSPMW was generally successful in maintaining the 40 cent base rate. In 1932, the union was willing to take a 5 per cent reduction, to 38 cents as the minimum rate.[91] In addition to keeping wages up at the 38 cent level, union working conditions and overtime provisions were generally maintained.

On June 4, 1932, President Burke wrote William Burnell, a vice-president:

> I think we have done remarkably well in view of present conditions. You are quite right in saying that it is wonderful that we have been able to come through so far with only a 5 per cent cut for our 40 cent men, in view of what is being paid common labor in most industries. In every town where I go I find common labor rates anywhere from 25 to 30 cents an hour, and in many cases much less. The Pulp, Sulphite Workers organization is practically alone in trying to hold up the labor rates.[92]

While the union was willing to accept moderate reductions, extraordinary cuts which could not be justified were met with resistance. On May 16, 1932, officials of the Union Bag and Paper Corporation proposed a wage reduction of 16⅔ per cent, elimination of the time and one-half premium for overtime, and introduction of the open shop. The union countered with an offer to reduce wages 10 per cent, this for a company that had a profit of over $100,000 in 1931.[93] When the company rejected the union counterproposal the union struck. Almost a year later, in June 1933, the walkout ended, with the company agreeing to recognize the union.[94]

---

89. W. Rupert Maclaurin, "Wages and Profits in the Paper Industry," *Quarterly Journal of Economics*, February 1944, pp. 204–205.
90. IBPSPMW, *Journal*, August 1932, p. 1.
91. *Ibid.*, p. 2.
92. John Burke to William Burnell, June 4, 1932. Papers of the International Brotherhood of Pulp, Sulphite and Paper Mill Workers, Wisconsin State Historical Society, Madison, Wis.
93. *Ibid.*, p. 3.
94. IBPSPMW, *Journal*, June 1933, p. 5.

Conditions in some nonunion mills were even worse. The Mersey Paper Company in Liverpool, Nova Scotia, cut its base rate to 27 cents.[95] In 1932, an unorganized mill in Quebec cut wages from 30 cents to 27 cents to 24 cents per hour in the six months from January to June.[96]

The depression was a severe blow to the unions. Membership of the Pulp Workers fell to 5,000 and the Paper Makers had only 4,000 members in 1933.[97] On July 22, 1932, President Burke wrote to International Auditor, George C. Brooks:

> I am very sorry to say that I shall not be able to send you a salary check for next week. I shall resume your salary payment in August. . . .
>
> Our receipts for July have been very, very poor, and we have had to pay out more than $700 to the Hudson Falls strikers. I am taking both Burnell and you off the payroll next Saturday and also two of our office staff.
>
> The mills are operating so poorly that it is becoming harder and harder to collect dues. However, we shall have to do the best we can, and I intend to do the best I possibly can for all those who have been faithful to our organization as long as possible.[98]

The depression of 1929 to 1933 reduced the influence of the paper industry unions almost to the vanishing point. Yet the next year was to see a resurgence of organization in the industry. At the end of 1967, approximately thirty years after the Great Depression, the IBPSPMW claimed more than 171,000 members in 723 locals, while the successor to the IBPM, the United Papermakers and Paperworkers, had 144,300 members in 731 locals.

## *The Upswing of Organization*

Passage of the National Industrial Recovery Act was one of the factors stimulating the resurgence of unionism in the paper industry. The labor provisions gave the unions "an unusual opportunity to organize the unorganized."[99] In addition to the guarantees for labor provided in the Act, the fact that labor was represented at the hearings concerning formulation of the Codes gave the union leaders a chance to meet the heads of companies. Both President Burke of the Pulp Workers and President Burns of the Paper Makers testified at the code hearings. Their moderate public statements, and their conversations with industry repre-

95. IBPSPMW, *Journal*, July 1931, p. 5.
96. IBPSPMW, *Journal*, August 1932, p. 3.
97. Leo Wolman, *Ebb and Flow in Trade Unionism* (New York, 1936), p. 183.
98. Papers, July 22, 1932.
99. IBPSPMW, *Journal*, June 1933, p. 3.

sentatives made a favorable impression. By the end of 1933, President Burke was able to report:

> The International Brotherhood of Pulp, Sulphite and Paper Mill Workers . . . is now enjoying an unprecedented growth in membership. New locals are being organized almost daily. The International Office has been literally swamped with work during the past few months. From all sections of the country—Pacific Northwest, Canada, Ohio, Virginia, Louisiana, Maine—come appeals for organizers.[100]

In August 1933, the presidents of the two unions in the paper industry had their first meeting since 1921 with the International Paper Company.[101]

Burke wasted no time in acting under NIRA. He spoke to officials of the Great Northern Company, who were participating in formation of the newsprint code in June 1933.[102] The codes provided an opening wedge for labor in the unorganized sectors. Burke and Burns were appointed labor advisors to the deputy administrator of the paper industry codes. Burke wrote to William H. Burnell on August 29, 1933:

> I returned from Washington Saturday morning from the hearing on the code of the American Pulp and Paper Association. There must have been about four hundred manufacturers at this hearing, practically all open shoppers. I think this hearing has broken down a great deal of opposition to our unions, because both Burns and I made a good impression at the hearing. If we use our heads now and do not make wild demands, I feel that we have an opportunity to organize the whole industry.[103]

By 1933, Pulp, Sulphite membership rose to 6,900.[104] The lean times were ending.

In 1933, too, President Burke could say something good about the union's old nemesis, the International Paper Company. Reporting on the hearings for the newsprint code, Burke noted he had met President Graustein of IP, a nonunion employer. "However, it should be said in all fairness to Mr. Graustein that the International Paper Company, under his leadership, has never taken . . . an anti-union stand."[105] The company permitted unions to function in the mills, though without written agreement.

In 1934, the Pulp, Sulphite membership rolls had increased to 8,500, while the IBPM had 11,500.[106]

---

100. IBPSPMW, *Journal*, December 1933, p. 2.
101. Papers, "Officers Report," August 1933.
102. *Ibid.*, "Officers Report," June 1933.
103. *Ibid.*, September 18, 1933.
104. Wolman, p. 183.
105. IBPSPMW, *Journal*, June 1933, p. 3.
106. Wolman, p. 183.

## *Organizing in New Areas—The South*

Prior to the 1930s, paper production in the South was insignificant. With the widespread application of the sulphate pulping process, however, the region's supply of fast-growing pine became a prime raw material. By 1943, 60 per cent of U.S. kraft production was located in the South.[107]

In 1918, an IBPM local had been organized at the Bogalusa Paper Company, Bogalusa, Louisiana (now part of Crown-Zellerbach). In order to smash the union, the company imported strikebreakers. After the death of several union men, the company succeeded in its efforts.[108] Unionism in the Southern pulp and paper industry was effectively halted until the late 1930s. At the convention in 1937, the decision was made to expand the organizing staff and place an organizer in the South. By the middle of 1938, the IBPSPMW had eight locals in the South.[109]

The breakthrough came in the South on January 18, 1938, when the Southern Kraft Corporation, subsidiary of the International Paper Company, signed an agreement recognizing the IBPSPMW and the IBPM as bargaining representatives for its employees. The contract also provided for maintenance of membership.[110] A base rate of 40 cents per hour was agreed upon. Organization of Southern Kraft gave the Pulp, Sulphite union seven locals at Bostrap, Louisiana; Camden, Arkansas; Mobile, Alabama; Moss Point, Mississippi; Panama City, Florida; Georgetown, South Carolina; and Springhill, Louisiana. Vice-President Sullivan of the IBPSPMW commented:

> Who says we cannot organize the South? Our organization has taken deep roots in the South. We find most of the Southern employers friendly and cooperative. We are building up some good wage rates in Southern Mills.[111]

By the end of World War II only a handful of firms in the South remained unorganized.[112]

## *The Lake States*

In the years following World War I and throughout the 1920s, the IBPSPMW maintained a presence in the Lake States. In 1931, President Burke was able to report five locals in Wisconsin and one in Minnesota.[113]

---

107. Brotslaw, p. 138.
108. *Ibid.*, p. 139.
109. IBPSPMW, *Journal*, May-June 1938, pp. 21–22.
110. IBPSPMW, *Journal*, March-April 1938, pp. 20–21.
111. IBPSPMW, *Journal*, July-August 1939, p. 4.
112. Brotslaw, p. 140.
113. IBPSPMW, *Journal*, April 1931, p. 14.

In 1931, agreement was reached at Hoberg Paper and Fibre Company in Green Bay, Wisconsin.[114] The other important union firm in Wisconsin was Consolidated Water Power and Paper Company of Wisconsin Rapids.

In 1933, President Burke sent Jacob Stephan, International representative, to Wisconsin to lead the organizing drive. Progress was slow, but, by the end of the year, Stephan was able to report some gains. Locals were established at Kaukauna, Combined Locks, Neenah, Menasha, Menominee, and Milwaukee. Members of the Neenah local worked in the mills of Kimberly-Clark, while the Menasha local was composed of Marathon employees. Stephan's work gave the union a total of eleven locals in Wisconsin.[115] He also journeyed to Michigan and helped re-establish Pulp, Sulphite locals at Escanaba and Manistique.[116] Stephan then turned over the organizing job to Ray Richards, who was to become a vice-president of the union. By the end of 1934, there were fifteen locals in Wisconsin, three in Michigan and three in Minnesota.[117]

The years 1935 to 1937 saw large scale organization in the Lake States. From April to September 1935, the union shop was received at Hoberg Paper and Fibre Company, Green Bay, Wisconsin; the Escanaba Paper Company, Escanaba, Michigan; and the Andawagan Paper Products Company, Wisconsin Rapids, Wisconsin. In 1935, too, agreement was reached with the Thilmany Paper Company, Kaukauna, the Combined Locks Paper Company, Combined Locks, and the Paper Supply Company, Menasha. The agreements with Thilmany and Combined Locks, in particular, represented significant gains for the union, since the agreement came in spite of the determined opposition of the companies. An agreement covering the workers at the Rothchild and Wausau plants of the Marathon Paper Company was signed on June 13, 1936.[118] A better fight against the union was waged by Mosinee Paper Mills Company. The NLRB ordered an election which the union won and a local was established. In late 1936 and early 1937, new locals were established at the Diana Paper Company, Green Bay; the Tomahawk Paper Company, Tomahawk; the Marathon Paper Company, Ashland; the Sterling Paper Company, Eau Claire, and the Peavy Paper Company, Ladysmith.[119] The organizing success in Wisconsin gave President Burke particular pleasure. He well knew the difficult task faced by the union, opposed as it was by the many industrial spies.[120]

114. IBPSPMW, *Journal*, July 1931, p. 2.
115. IBPSPMW, *Journal*, December 1933, p. 3.
116. *Ibid.*
117. IBPSPMW, *Journal*, October 1939, pp. 12–13.
118. IBPSPMW, *Proceedings of the Seventeenth Convention*, pp. 145–147.
119. *Ibid.*
120. *Ibid.*

Burke had been in Wisconsin during an organizing campaign in 1916 —a campaign that failed. He recalled a strike at Interlake Paper Company in Appleton:

> In that Interlake strike in 1916 I remember distinctly the brutality of the police in Appleton. I remember distinctly one morning on the picket line when the police ran amuck and almost killed one of our strikers . . . and they almost knocked me unconscious. I got a terrible blow back of the ear, which knocked me ten or twelve feet. I remember how our boys were arrested and the thugs employed by the company took over the police department in Appleton and we had no support from anybody. We were alone, and had to fight these battles against all of these powers. Now, we have a report of all this progress. We meet all these companies, we have agreements with all these companies, and it is only a matter of time until we have agreements with all the companies in Wisconsin.[121]

## *Other Organizing Success*

Perhaps the most conspicuous accomplishment of the unions during the second half of the 1930s was the organization of the International Paper Company. (Organization of the Southern Kraft Division has been discussed earlier.) In April 1936, Pulp Workers Vice-President Maurice LaBelle journeyed to Temiskaming, Quebec, a company town owned by International Paper. The mill there had been recently purchased by IP from the Riordan Company, but local management which discouraged union organization had been retained. Men were fired because they favored the union, and organizers were run out of town or forced to leave because they could not find a place to stay at night.[122] LaBelle, however, was successful in holding meetings in the middle of the Canadian winter, and management responded with its traditional answer to organization, the company union. By then, however, it was too late. Wage increases were negotiated that raised the base rate from 25 cents to 34 cents per hour; time and one-half for overtime on Sundays was granted, as well as premium payment for work over ten hours per day.[123] In 1937, President Burke reported that the Canadian International Paper Company (IP's Canadian subsidiary), with mills in Gatineau, Three Rivers, and Temiskaming, Quebec, Dalhousie, New Brunswick, and Corner Brook, Newfoundland, had reached agreement with the union.[124] This agreement was of major significance to the unions in the paper industry. It marked the resumption of contractual relations with the industry leader, whose very size and significance gave it great influence on the actions of other firms.

---

121. *Ibid.*, p. 150.
122. *Ibid.*, p. 47.
123. IBPSPMW, *Journal*, April 1936, p. 6.
124. IBPSPMW, *Journal*, May-June 1937, p. 1.

With the agreements at Southern Kraft and Canadian IP, there remained only one segment of International Paper Company unorganized, the Book and Bond Division. It was the book and bond mills of Fort Edward and Palmer, New York, and Livermore Falls, Maine, that were so significant in the 1921 strike against IP. In September 1937, a conference was held in South Glens Falls, New York, between representatives of the two unions and the Book and Bond Division of IP. While the meeting adjourned for thirty days with no action being taken, delegates reported the conference was conducted in an agreeable fashion.[125]

On October 28, 1937, at a meeting in South Glens Falls, agreement was reached calling for a union shop at IP mills in Fort Edward and Palmer, New York, and Livermore Falls, Maine. The agreement covered approximately 2,500 employees. Shortly thereafter an agreement was signed covering the 700 employees of the IP mill at Niagara Falls.[126]

By the middle of 1938, the Pulp, Sulphite Workers had 238 local unions from Maine to California, Minnesota to Arkansas, and from Newfoundland to British Columbia in Canada.[127]

## *The Rise of the CIO*

During the early years of the CIO, while its main energies were concentrated on the auto and steel organizing campaigns, workers in several paper and converting plants were swept into the ranks of the new organization. These workers were temporarily put into local industrial unions. At the first CIO Convention in 1938, representatives of the paper industry local industrial unions met for the first time, and agreed to work for the formation of their own international union.[128]

The CIO had chartered the Paper, Novelty and Toy Workers International Union in 1938. Its initial membership base was somewhat under 5,000,[129] concentrated mainly in smaller plants, and generally not in the mills, where the AFL unions had preceded it. As the number of local industrial unions in the paper industry increased, the leaders of the CIO recognized the desirability of granting increased status to the membership. In 1940, the paper industry local industrial unions were combined with the Paper, Novelty and Toy Workers.[130] The name of the organization was changed to the United Paper, Novelty and Toymakers International Union, and the union made moderate organizational gains. In 1940

125. IBPSPMW, *Journal*, September-October 1937, p. 24.
126. IBPSPMW, *Journal*, November-December 1937, p. 1.
127. IBPSPMW, *Journal*, May-June 1938, pp. 21–22.
128. United Paperworkers of America, *Officers Report to the Fourth Constitutional Convention*, April 25–29, 1955, p. 13.
129. Leo Troy, *Trade Union Membership*, 1897–1962 (New York, 1965), pp. A22–A26.
130. UPA, *Officers Report*, p. 13.

membership was approximately 9,100; by 1943 it was close to 20,000.[131]

While membership was increasing, internal division within the union was also on the rise. The combination of toy workers and papermakers provided little basis for lasting unity, as the members of both groups sought their own identity and recognition. The paper workers were led by Harry D. Sayre, who was later to rise to high office in the United Papermakers and Paperworkers. The toy workers followed Anthony Esposito, the union president.[132] Following the 1942 convention, the two groups made the decision to go their separate ways and on January 1, 1944, Philip Murray granted a charter to the Paper Workers Organizing Committee.[133]

Headquarters of the PWOC were in Cincinnati, Ohio. CIO President Murray appointed Allan S. Haywood as chairman of PWOC. Robert J. Davidson was director, with Harry Sayre secretary-treasurer. Members of the executive committee were Charles Bridgewater, James Carey, Frank Grasso, Burt Mason, David McDonald, and R. J. Thomas.[134] The PWOC took from the Toy and Novelty Workers 78 local unions, 12,000 members, $2,000 in cash, $500 worth of office supplies and equipment, and a staff of fourteen.[135]

In July 1944, Davidson was succeeded as PWOC director by Walter Smethurst, and in January the headquarters was moved to Cleveland.[136] Progress in the early months and years was fairly rapid. By 1946 the union had over 24,000 members, and two years later its ranks were in excess of 30,000.[137] In November 1945, Director Smethurst resigned because of poor health, and Murray appointed Frank Grasso to succeed him.

The PWOC in 1944 scored one of its significant victories by winning bargaining rights from the IBPM and the Pulp, Sulphite Workers in the mills of the West Virginia Pulp and Paper Company at Covington, Virginia, Luke, Maryland, and Williamsburg, Pennsylvania. The mills employed 4,000 workers, and the victory gave the PWOC its first significant foothold in basic pulp and paper. Prior to its success at West Virginia Pulp and Paper, the PWOC had been almost exclusively confined to converting plants.[138]

The PWOC officially changed its name in 1946 to the United Paperworkers of America, and in 1947, the first policy convention was held in Cleveland. In October of that year, Philip Murray named Harry Sayre,

131. Troy, pp. A22–A26.
132. Brotslaw, p. 158.
133. UPA, *Officers Report*, p. 13.
134. *Ibid.*
135. *Ibid.*
136. *Ibid.*
137. Troy, pp. A22–A26.
138. Brotslaw, p. 160.

*Membership in Pulp and Paper Industry Unions*
*(Thousands)*

| | Paper Makers | Pulp Workers | Paper Workers | Paper, Novelty and Toy Workers |
|---|---|---|---|---|
| 1935 | 9.3 | 15.7 | | |
| 1936 | 7.7 | 17.0 | | |
| 1937 | 17.1 | 30.0 | | |
| 1938 | 23.0 | 37.9 | | 4.8 |
| 1939 | 26.7 | 43.5 | | 2.5 |
| 1940 | 31.0 | 51.9 | | 9.1 |
| 1941 | 36.9 | 60.0 | | 15.7 |
| 1942 | 39.0 | 64.0 | | 15.7 |
| 1943 | 39.9 | 68.0 | | 19.7 |
| 1944 | 43.5 | 73.9 | 14.0 | |
| 1945 | 47.9 | 80.0 | 18.6 | |
| 1946 | 51.2 | 97.7 | 24.3 | |
| 1947 | 56.0 | 118.3 | 32.6 | |
| 1948 | 56.5 | 120.4 | 30.8 | |
| 1949 | 59.8 | 123.7 | 31.2 | |
| 1950 | 64.1 | 130.3 | 32.0 | |
| 1951 | 67.6 | 141.6 | 35.9 | |
| 1952 | 67.4 | 139.4 | 34.3 | |
| 1953 | 70.2 | 148.9 | 39.0 | |
| 1954 | 72.8 | 149.9 | 39.9 | |
| 1955 | 75.2 | 160.1 | 37.5 | |
| 1956 | 80.3 | 165.8 | 38.7 | |
| 1957 | 118.3 | 162.2 | | |
| 1958 | 121.6 | 165.4 | | |
| 1959 | 126.3 | 170.4 | | |
| 1960 | 129.4 | 170.5 | | |
| 1961 | 126.2 | 171.1 | | |
| 1962 | 130.6 | 174.1 | | |

Source: Leo Troy, *Trade Union Membership, 1897–1962,* National Bureau of Economic Research (New York, 1965), pp. A6–A19.

president and Frank Grasso, secretary-treasurer of the UPA, posts they were to hold until the merger with the Papermakers in 1957.

The challenge of the CIO did not go unnoticed by the International Brotherhood of Pulp, Sulphite and Paper Mill Workers. They stood to lose by the formation of a rival group attempting to organize the unskilled and semiskilled workers along industrial lines. At the 1939 convention, a resolution was passed calling for a study of the wages, hours and working conditions of the converting section of the industry.[139] Two years later, at the 1941 convention a report was made on the results of the study. Pulp, Sulphite Vice-President John Sherman, who conducted the study found that the principal AFL union having jurisdiction in the industry was the Pulp Workers, although the Printing Press-

139. IBPSPMW, *Proceedings of the Eighteenth Convention,* March 14–17, 1939, p. 142.

men had some locals covering specialty paper work having a direct relationship to printing.[140] Sherman noted the existence of the CIO Toy and Novelty Workers. The report concluded:

> The Converted Paper Products Industry is one of the [sic] vital importance to this International Union; first, because manufacturers of pulp and paper are branching out into the industry; secondly, because the Pulp and Paper Industry is vulnerable to attack as other unions are making inroads from the Converting side, knowing full well that in the Pulp and Paper Industry wages have been increased to such an extent that the workers on the average are fairly well satisfied with the conditions.[141]

The UPA attempted to organize aggressively, but it was faced with a set of circumstances beyond its control. The two AFL unions had been in the field for many years prior to the formation of the UPA, and they had battled the companies and had come to terms with most of them. By the time the UPA came on the scene the bulk of the basic pulp and paper industry had been organized by the IBPM and the Pulp, Sulphite Workers.

An additional handicap faced by the UPA was its association with the "radical" CIO. It may be possible that some firms preferred to deal with the AFL unions, which were led by men who had demonstrated a capacity to see the problems of the industry sympathetically rather than with a CIO union whose leadership was an unknown quality. The emergence of the CIO as a force in paper industry unionism saw the start of a period of ferment as the CIO raided the AFL and the AFL reacted in kind.

## *Other Rival Unions*

The UPA was not the only organization to be sniping at the AFL unions during the 1940s and 1950s. District 50 of the United Mine Workers and the International Woodworkers Association also attempted to organize in the pulp and paper industry. In 1938, four federal locals of pulp and paper workers had sought and obtained membership in District 50 of the UMW.[142] District 50 campaigned intensively against the AFL unions, particularly in the Northeast. Their efforts achieved some success, as they won bargaining rights at the Oxford Paper Company in Rumford, Maine, the Mechanicsville, New York, plant of West Virginia Pulp and Paper, the Niagara Falls plant of Kimberly-Clark, and the Fort Edward, New York, mill of Scott Paper Company.[143] The AFL

140. IBPSPMW, *Proceedings of the Nineteenth Convention*, September 8–12, 1941, p. 212.
141. *Ibid.*, p. 213.
142. Brotslaw, p. 164.
143. *Ibid.*

unions attempted to retaliate. District 50 had organized the Berlin, New Hampshire, mill of the Brown Company, which at that time employed about 2,700 men. Some dissatisfaction with the UMW developed when they left the CIO, became independent, affiliated with the AFL and then left again. Things were happening too fast for the men at Berlin. There was also unrest over the lack of representation in the councils of District 50. A delegation visited John L. Lewis to present their grievances, but he gave them a hostile reception. The UMW conduct of a three-week strike, in which the men felt they were being undermined by the UMW district director, was the issue that caused a break. A representative of the local visited Fort Edward and received a charter from John P. Burke. In the ensuing NLRB election, the UMW got less than 300 votes out of the 2,700 eligibles.[144]

President Burke of the Pulp Workers expressed himself strongly in 1950 at the convention on the subject of rival unions:

> Our International Union accepts the challenge of the UPA of the CIO. We accept the challenge of every rival union. This Union will go ahead and make progress in spite of all their opposition and all of their treachery, and all of their deceit, and all of their immoral and unmoral methods. They haven't destroyed us; they are not going to destroy us.[145]

Thus Burke accepted the external challenge of "every rival union" with the bold statement that "they are not going to destroy us," but the internal dissension to come on the West Coast was to be a different matter where challenge alone would not suffice to hold the ranks of the International.

---

144. IBPSPMW, *Proceedings of the Twenty-Second Convention,* August 10–19, 1950, pp. 77–79.
145. *Ibid.,* p. 83.

# 2

# ORGANIZATION ON THE WEST COAST

Prior to the great growth of the unions in the paper industry in the 1930s, membership was confined almost exclusively to New England and Eastern Canada. The unions continued to maintain a presence in the Lake States through the 1920s, but on the West Coast their rudimentary organization disappeared.

Before 1920, the Pulp, Sulphite Workers had several local unions in the Pacific Northwest. In 1918, however, the locals lost a strike against the Crown-Willamette interests. The loss of the Crown strike effectively destroyed the unions on the coast, and not even a token presence was maintained through the 1920s. The firms in the region were generally antiunion and this, coupled with the distance from headquarters across the continent at Fort Edward, may have hindered organization. When the depression of the 1930s developed, the unions were on the defensive. unable to even consider expanding into the Pacific Northwest. Problems closer to home seemed to have greater urgency.

President Burke was able to devote some thought in mid-1933 to organization in the Northwest, and on June 27 he wrote to William Green, president of the American Federation of Labor:

> There are a number of unorganized pulp and paper mills in the states of Oregon and Washington. For some reason or other the paper mill unions have not been able to make much progress on the Pacific Coast. However, I think we might have a chance to break into that field at the present time if we were only in a position to send organizers out there. Unfortunately, we are not in a position at this time to finance sending organizers to the Pacific Coast.
>
> Therefore, I should appreciate it greatly if you would write a letter to the officers of the State Federation of Labor of Washington and Oregon, also to any organizers of the American Federation of Labor who may be in that district, asking if they will give some attention to organizing the workers in the mills of the Crown-Willamette Paper Company . . . and the mills of the Fibre Board Products Inc. . . . and the National Paper Products Company . . . .
>
> I would appreciate it if the officers of the Washington State Federation of Labor or any American Federation of Labor organizer in that district will make a special attempt to organize the mill of the Weyerhauser Timber Company at Longview, Washington. The general man-

> ager of this mill is Mr. Robert B. Wolf who was formerly manager of the Spanish River Pulp and Paper Company and who is very friendly to organized labor.
>
> Mr. Wolf is now in the East. Within the past few days I have had a telephone conversation with him. He is very much opposed to the formation of company unions . . . . The concern for which Mr. Wolf works, the Weyerhauser Timber Company, is a very powerful concern, but if we can get an organization started at Longview it would give Mr. Wolf something with which to work to break down the prejudice of his concern against union labor.[1]

Shortly thereafter, President Green took action and reported to President Burke:

> Most assuredly I will be glad to cooperate in every way possible in trying to organize the paper mills on the Pacific Coast. With that purpose in view, I will immediately communicate with our Organizer for the Northwest, C. O. Young in Tacoma, and also through the Secretary of the Washington State Federation of Labor and the Oregon State Federation of Labor, giving them a copy of your letter and asking their aid and cooperation in organizing the pulp and sulphite workers in the several mills which you list.[2]

Green did indeed communicate with C. O. Young, and a month later Young wrote Burke requesting a charter for a Pulp, Sulphite local at Longview, Washington.[3] The local was started with thirty-eight members, but Young was optimistic about prospects for enrolling the entire work force there.[4]

Throughout the remainder of 1933 organization proceeded in the Northwest. In September, Gladys Funkhauser of the Grays Harbor County Central Labor Union reported to Burke the organization of a local at Hoquiam, Washington, with a membership of eighteen.[5] In August Ben T. Osborne, executive secretary-treasurer of the Oregon State Federation of Labor, wrote Burke offering his services as an organizer for the Pulp, Sulphite Workers, and in late September established Oregon City Local 68, at Oregon City, Oregon.[6] Other locals were set up at St. Helens and Salem, Oregon, and Shelton, Sumner, Tacoma, Port Angeles, and Vancouver, Washington, by the end of October, 1933.[7]

In the late summer of 1933, pressure developed from the new locals for a visit from a representative of the organization to which they belonged. George J. Schneider, a vice-president of the Paper Makers and a member

---

1. John Burke to William Green, June 27, 1933. Papers of the International Brotherhood of Pulp, Sulphite and Paper Mill Workers, Wisconsin State Historical Society, Madison, Wis.
2. Papers, William Green to John Burke, July 3, 1933.
3. *Ibid.*, C. O. Young to John Burke, August 3, 1933.
4. *Ibid.*
5. *Ibid.*, Gladys Funkhauser to John Burke, September 26, 1933.
6. *Ibid.*, Ben T. Osborne to John Burke, September 22, 1933.
7. *Ibid.*, "Papers of Longview, Washington, Local 153."

of Congress, was sent from Wisconsin in the fall to coordinate activities on the Coast.[8] President Burke sent a circular letter to all locals in the Northwest informing them of Schneider's impending visit:

> We are sending Brother Schneider to the Northwest at this time because we understand there is a movement on foot to have a conference of delegates from the locals at some central point. We think Brother Schneider can be of great assistance in giving advice and guidance.[9]

The conference referred to in Burke's letter was the idea of Ben Osborne. On October 1, Osborne wrote Burke and Arthur Huggins of the IBPM:

> This Council wants these two unions to thrive and prosper and, of course, to do that they must get results for their members. That is why we are concerned about the growth of your organizations in the Pacific Coast.
>
> I have, therefore, suggested that a conference of organized unions and leaders at unorganized points be held very soon at some central point, say Longview, Washington or Portland.
>
> The President of Capital Local 230, Mr. Charles E. Davis . . . seems to have considerable understanding of the work before these organizations.
>
> For convenience and to facilitate matters, I suggest each of you write Mr. Davis and authorize him to call a conference at Longview, Washington . . . .[10]

At the end of October, a meeting of all Pulp, Sulphite and Paper Maker locals in the Northwest was held on the call of Capital Local 230 and the Salem Trades and Labor Council. Fifty-four delegates from ten locals were in attendance.[11] The conference heard reports on organizing and the impending visit of George Schneider. E. P. Fourre of Local 161 was elected president and John Sherman of Local 155 secretary-treasurer. Before adjournment, Fourre appointed a committee on permanent organization and by-laws. The locals meeting at Longview called themselves the Pacific Northwest Conference of Pulp and Paper Employees.[12] This conference marked the formation of a regional group within the Pulp Workers, and it was to proceed to develop its own intrastructure, including per-capita tax, constitution, and bargaining philosophy.[13]

---

8. *Ibid.*, John Burke to R. S. MacDonald, October 26, 1933.
9. *Ibid.*, Circular letter to all locals in the Northwest, October 25, 1933.
10. *Ibid.*, Ben T. Osborne to John Burke, October 1, 1933.
11. *Ibid.*, "Minutes" of the Pacific Northwest Conference of Pulp and Paper Employees, Longview, Wash., October 29, 1933.
12. *Ibid.*, "Minutes."
13. In the judgment of the author, it was the existence of the Pulp and Paper Employees Association (as the Conference came to be called) that was a necessary condition for revolt within the International.

## *Visit of George Schneider*

The trip of George Schneider of the Paper Makers marked a shift in the organization campaign on the West Coast. The presence of Schneider marked the first introduction of the International hierarchy, and Schneider concentrated his efforts on reaching agreements with the firms of the region. In their attempts to come to terms with the companies, the unions had a friend in the person of Robert B. Wolf, manager of the Weyerhauser Timber Company—Pulp Division at Longview, Washington. Before moving to Weyerhauser, Wolf had been general manager of the Spanish River Pulp and Paper Company in Canada. Spanish River had mills at Sault Ste. Marie, Espanola, and Sturgeon Falls, Ontario. While there, Wolf had become acquainted with the unions and had come to accept them. He had granted a union shop to the unions at the mills of the Spanish River firm, and had been able to live with them. When he moved to Weyerhauser, his attitude was an opening wedge to contact with the management of the companies of the Coast.

As Schneider was preparing to leave for the Coast, Burke wrote:

> Robert B. Wolf is the manager of the Weyerhauser Timber Company of Longview, Washington. I met Mr. Wolf in New York one day last week and had a short visit with him. He told me that several of the companies out there are rather friendly to the unions.[14]

Schneider arrived on the West Coast late in 1933 and began attempting to contact the firms in the area, but all did not go smoothly. He found resistance in an unexpected quarter, as he told Burke on December 3:

> I met Bob Wolf at Longview Friday.
>
> He is not ready to sign an agreement. He raises the question as to whether [sic] employers can sign agreement with unions forcing men to join.
>
> The Crown-Willamette Company is opposed to organization.[15]

A few days later, however, Schneider was able to report some progress to Burke by telegram:

> Conferences with George Berkey [sic] Manager Crown-Willamette located here resulted in clearing situation some. Can meet Block in San Francisco on way home.[16]

The record from this point indicates that Schneider was successful in converting Burkey of Crown to the union cause. It was Burkey who assisted the unions throughout the remainder of 1933 and during the next

14. Papers, John Burke to George Schneider, October 25, 1933.
15. *Ibid.*, George Schneider to John Burke, December 3, 1933.
16. *Ibid.*, Telegram from George Schneider to John Burke, December 6, 1933.

year, and his attitude was instrumental in setting up the historic conference in August 1934. The record continues with the following letter from Schneider to Burke on December 16, 1933:

> I am leaving here tomorrow . . . for San Francisco where I have an appointment with Mr. Block, President of the Crown-Willamette Company. I will also try to contact Mr. J. D. Zellerbach of the Crown-Zellerbach Corporation. I have had several meetings here (Portland, Ore.) with Mr. George Burkey, manager of the several mills in this section of the country and he arranged the appointment for me.[17]

After his meeting in San Francisco Schneider left for home.

For the first three months of 1934, the Internationals did not have an experienced organizer in the district. On April 12, President Burke appointed Pulp, Sulphite Vice-President Herbert Sullivan to represent the unions on the coast and to conduct organizing and negotiating work. Robert Wolf assisted Sullivan's organizing efforts, helping him make contact with the management of firms in the region. Sullivan reported:

> I had another conference with Bob Wolf Friday and he thought it a good plan to have the locals submit their union shop agreement and wage scale. He also agreed with me that I should meet as many of the mfgrs. as I can and he will tell them to expect me.[18]

Sullivan continued to maintain relations with George Burkey, manager of the Crown interests. With Burkey's help, contact had been made with top management of the Crown complex by Schneider the previous December. In May 1934. Sullivan reported to Burke on his discussions with Burkey.

> Geo. P. Burkey who is the big shot in paper in this district proposed that a conference of all employers in the district be arranged and that all locals send about 2 delegates to see if we could not reach an agreement . . . . He thought that such a conference could be arranged and would try and arrange it if we can only keep these fellows in line till such a meeting can be arranged. I think we can put over the first stage leading up to the union shop but the dam [sic] idiots can't seem to get it into their heads that there is a long future in this for them if it is handled right. At this time some of them think they must fight to get what they are after and the Longshoremen's strike is getting everyone in a fighting mood.[19]

This appears to be the initial move towards the meeting of August that year which marked the beginning of contractual relations with the employers. However, before a meeting could be held, the record indicates many false starts and difficulties. The complex of the Crown-Zellerbach and Crown-Willamette firms were the largest in the area, and their participation in discussions with the unions was critical in the union drive

17. *Ibid.*, George Schneider to John Burke, December 16, 1933.
18. *Ibid.*, Herbert Sullivan to John Burke, April 29, 1934.
19. *Ibid.*, Herbert Sullivan to John Burke, May 24, 1934.

for acceptance. Evidently, the top management of the Crown empire vacillated somewhat before deciding to deal with the unions. On June 16, Sullivan telegraphed Burke: "Have been advised to call on Crown-Zellerbach officials in San Francisco will leave for there Sunday night."[20] A letter of the same date amplified the telegram:

> I have had interviews with all the mill mgrs. in the district and they seemed to disagree with the idea but they did not say they would not attend the conference. I had a conference with Mr. Raymond in Seattle who is the big shot over the mills in the Olympia district and he advised that I go to San Francisco and take the matter up with Messrs. Block, Mills and Zellerbach. I talked it over with Bob Wolf and he thought I should . . . .[21]

Evidently Sullivan's initial reception at Crown-Zellerbach headquarters was lukewarm, but the managers changed their attitude after several days and Sullivan reported:

> I stopped off here (S.F.) for another meeting with the Crown Zellerbach Co. At the meeting with them last week they did not seem to think so well of a general conference of all the employers in the N.W. District but today agreed to go along with the program the meeting to be arranged as soon as there is an understanding as to what the code will be.[22]

By the early part of July, however, the Crown management seemed to be having second thoughts. George Burkey expressed doubt that C-Z would go through with the conference.[23] But on July 13, Sullivan reported to Burke there had been another change of heart on the part of the employers, and that arrangements were made to hold a conference. George Burkey of Crown-Willamette seems to have been instrumental in securing employer acceptance for the idea of a conference.

> Talked with Burkey today he says all employers in district will attend joint conference he had meeting of the employers in Portland District Tuesday all agreed you or Burns or both should attend this first conference. J. D. Zellerbach will attend.[24]

Ten days later, Sullivan telegraphed Burke with the final arrangements for the conference in Portland:

> Conference arranged for Wednesday August first at Nine Thirty AM. Arrange to be here Tuesday July Thirty First will hold meeting of delegates Tuesday afternoon.[25]

---

20. *Ibid.*, Telegram from Herbert Sullivan to John Burke, June 16, 1934.
21. *Ibid.*, Herbert Sullivan to John Burke, June 16, 1934.
22. *Ibid.*, Herbert Sullivan to John Burke, June 28, 1934.
23. *Ibid.*, Herbert Sullivan to John Burke, July 9, 1933.
24. *Ibid.*, Telegram from Herbert Sullivan to John Burke, July 13, 1934.
25. *Ibid.*, Telegram from Herbert Sullivan to John Burke, July 23, 1934.

## *Factors Favoring Acceptance of the Unions*

There were a number of factors operating in behalf of the unions in their drive to organize the industry in the Pacific Northwest. One of the most significant was perhaps the employers' fear they would be organized by other, perhaps radical labor unions.

The years 1933–34 were marked by great labor struggles on the West Coast. In San Francisco, during the fall of 1933 there was a surge of interest in the International Longshoreman's Association. By the spring of 1934 negotiations for a contract had come to a standstill, and on May 9 longshoremen from Puget Sound to San Diego went on strike. The strike spread, and Teamsters boycotted the docks. Other maritime unions representing Sailors, Marine Firemen, Marine Cooks and Stewards, and licensed officers declared strikes. In June, the Teamsters voted not to handle freight that had been worked by strikebreakers. Their action crippled the ports. When the employers attempted to open the port of San Francisco on July 3, fighting flared between pickets and the police and strikebreakers. On July 5, fighting resumed, police fired into a crowd and two people were killed. The unions voted a general strike and on July 16, city life slowed to a crawl. Vigilante committees were formed, strike leaders were accused of being "radicals" and "communists," and of leading a revolution. Police raided "radical" headquarters and hundreds were arrested. Strike restrictions were eased on the second and third days, and on the fourth day the general strike ended in San Francisco.[26] Seattle too was a battleground during the spring and summer of 1934. Mayor John Dare denounced the strike as "a soviet of longshoremen."[27] During the strike, police went to the waterfront and dispersed the strikers with tear gas. At the end of June, the head of the Seattle "citizens committee" appealed to the police to "run about 200 reds and aliens, who are responsible for a reign of terror, out of town."[28] On July 29, the West Coast longshoremen began to return to work and an arbitration award handed down on October 12 provided for a coast-wide bargaining unit.[29]

There is some indication that the violence of the longshore situation, with its attendant hysteria over "communists," "radicals," and "aliens," may have helped the unions in the paper industry secure acceptance from the employers. On May 5, 1934, Sullivan wrote Burke:

---

26. David F. Selvin, *Sky Full of Storm: A Brief History of California Labor* (Berkeley, Calif., 1966), p. 48.
27. Charles P. Larrowe, *Shape-Up and Hiring Hall* (Berkeley, Calif., 1965), p. 97.
28. *Ibid.*, p. 101.
29. *Ibid.*, p. 103.

> All of them [the mill managers] speak of the Communists . . . and I think they may be beginning to lean a little to the trade unions on that account but like the past when it is good for them they accept it.[30]

The firms of the region were evidently conscious of the tense situation, and took no action to provoke the fledgling paper unions. At the end of June, Sullivan wrote:

> The longshore strike is still the big issue. Several of the paper mills are shut down as they can't get any fuel oil to run. The Co.'s are being very careful not to aggrevate the situation by bringing in oil that has been handled by scabs or carried on ships manned by scabs.[31]

There is evidence to indicate that Burke adopted a policy of urging restraint upon the locals in the Northwest. The experience of the 1920s was still fresh in the minds of the officers of the International, and Burke was not about to throw away the opportunity for organization of the region by the inopportune action of a few locals. He admonished Sullivan:

> Barnes and you should take a strong stand and not permit any of our locals to participate in a general strike without the sanction of any of the Executive Board.[32]

Burke's reasonable attitude and his demonstrated desire to keep friction to a minimum were probably significant influences on the firms of the region.[33]

Another factor operating in the union's favor was the passage of the National Industrial Recovery Act. The newsprint and pulp and paper industries were to formulate codes under NRA, and the hearings in Washington on the content of the codes provided significant contacts for the unions. Through the second half of 1933, Burke and Matt Burns of the Paper Makers journeyed frequently to Washington to participate in code hearings. In addition, the union leaders were often in attendance at meetings of various trade associations in the industry. The employer chairman for the newsprint board was President Whitcomb of the Great Northern Paper Company. The unions had maintained relations with Great Northern since the early years of the century, and Burke was able to comment:

> I think this will give the unions a good break, because, of course, the Great Northern is very friendly to us.[34]

30. Papers, Herbert Sullivan to John Burke, May 5, 1934.
31. *Ibid.*, Herbert Sullivan to John Burke, June 28, 1934.
32. *Ibid.*, John Burke to Herbert Sullivan, June 9, 1934.
33. In an interview at Fort Edward, N.Y., on September 20, 1966, Francis Tierney, office manager of the IBPSPMW told the author that in his opinion the longshore strike and the general strike in the spring and summer of 1934 were definitely factors influencing the companies in their dealings with the unions.
34. Papers, John Burke to George C. Brooks, June 10, 1933.

By the middle of October, Burke seems to have been accepted by the leaders of the industry. On October 15, he attended a meeting of the Executive Committee of the American Paper and Pulp Association.[35]

With the performance of Burke and Burns at the hearings for the NRA codes and their appearance at various trade association meetings and in informal conversations with management officials, much hostility towards the unions seems to have been overcome. Burke reflected this sentiment when he wrote:

> I returned from Washington Saturday morning from the hearing on the code of the American Pulp and Paper Association [sic]. There must have been about four hundred manufacturers at this hearing, practically all open-shoppers. I think this hearing has broken down a great deal of opposition to our unions, because both Burns and I made a good impression at the hearing. Huggins [vice-president of the Paper Makers] also did very well at the hearing September 14. If we use our heads now and do not make wild demands, I feel that we have an opportunity to organize the whole industry.[36]

## *Actions of the Pacific Northwest Pulp and Paper Mill Employees Association*

At the same time the hierarchy of the Internationals was endeavoring to convince the companies to adopt a reasonable attitude, the Employees Association was establishing itself within the union. On January 24, 1934, the Association met in convention in Port Angeles, Washington. At this meeting a resolution was passed imposing a per capita tax of 2 cents and a special assessment of $2.00.[37] Another meeting was held at Vancouver, Washington, on April 19, and a wage scale committee was formed looking to the initiation of bargaining. The report of the meeting noted: "it was moved at this time that each local take the scale and contract up with their employers. . . ."[38] Sullivan, in the meantime, was busy attempting to arrange a conference with the employers, which was ultimately held on August 1–3, 1934.

35. *Ibid.*, John P. Burke "Activity Report" for October, 1933.

36. *Ibid.*, John Burke to William H. Burnell, September 18, 1933.

37. *Ibid.*, "Minutes" of the Convention of the Pacific Northwest Pulp and Paper Employees Association, January 24, 1934. A history of the Employees Association incorporated in the minutes of the Association for January 1958 makes no mention of the meeting of January 24, 1934. The history states the per capita tax and assessment were voted on April 19, 1934.

38. *Ibid.*, "Minutes" of the Pacific Northwest Pulp and Paper Employees Association, April 19, 1934. This would indicate that the idea of a Uniform Labor Agreement, signed in August 1934 had not yet been discussed. However, former Pulp Sulphite Vice-President John Sherman, who was secretary of the Association at this time states in his book, *Twenty Years of Collective Bargaining and Twenty Years of Peace*, that at the April meeting the idea of a Uniform Labor Agreement was suggested by Vice-President Sullivan and accepted by the delegates. This author finds no record of any suggestion for a ULA in the minutes of the Employees Association.

On July 31, delegates from nineteen Pulp, Sulphite and Paper Maker locals met at the Multnomah Hotel, Portland, Oregon. They elected a bargaining board of six, including John Sherman, who later became a vice-president of the Pulp Workers.[39] On August 1, the first bargaining session began. John Burke headed the union side, and J. D. Zellerbach of Crown-Zellerbach led the employers. After three days of discussions an agreement was reached covering the two unions and fourteen companies.[40] A recognition clause was drafted providing for the Paper Makers and the Pulp Workers as agents for the purpose of collective bargaining under Section 7a of the NRA. Wages were increased to 45 cents for men and 37 cents for women. Three holidays, the Fourth of July, Labor Day, and Christmas were granted. A joint committee was established to determine the prevailing wage for various jobs, and mills paying below the prevailing rate were to raise their pay scales. The other significant contractual provision established a grievance system terminating in a Joint Relations Board.[41]

It is not clear who originated the idea for the Uniform Labor Agreement. Vice-President Sherman of the Pulp Workers reports the concept came from Herbert Sullivan of the Pulp Workers,[42] but no evidence seems to support the idea that the ULA originated with the unions. On the contrary, there is some evidence to indicate the initiative came from the companies. Officials of the Pulp, Sulphite Union told this writer that they believe management suggested the idea.[43]

## *Subsequent Developments*

In 1935, the procedure established the preceding year was followed. Twenty-six local unions attended the wage conference that began June 1. An increase of 2½ cents for men and 1 cent for women was secured, raising rates to 47½ cents and 38 cents respectively. In 1936, however, the harmony that prevailed in the previous period was strained. Pulp, Sulphite Local 169 of Hoquiam, Washington, voted to leave the joint

---

39. *Ibid.*, "Minutes" of the Pacific Coast Pulp and Paper Mill Employees Association, January 1958, p. 5.

40. The firms involved were: Weyerhauser Timber Co.—Pulp Division, Longview Fiber Co., Pacific Strawboard and Paper Co., St. Helens Pulp and Paper Co., National Paper Products Co., Columbia River Paper Mills Co., Oregon Pulp and Paper Co., Hawley Pulp and Paper Co., Crown-Willamette Paper Co., Rainier Pulp and Paper Co., Washington Pulp and Paper Corp., Olympic Forest Products Co., Fibreboard Products Inc., and Grays Harbor Pulp and Paper Co.

41. "Report of Organizer Frank C. Barnes," *Pulp, Sulphite and Paper Mill Workers Journal*, October 1934, p. 3.

42. John Sherman, *Twenty Years of Collective Bargaining and Twenty Years of Peace*, (Privately printed in Glens Falls, N.Y., 1954), p. 8.

43. Interview with Francis Tierney, Fort Edward, N.Y., September 20, 1966. Unfortunately, most of the participants in the original 1934 conference are no longer alive to supply information on this matter.

wage conference, believing that they could get a better settlement by negotiating independently. John Sherman, then a Pulp Worker vice-president wrote to President Burke:

> This matter, I think, should be given considerable thought, for I realize that all wage agreements and contracts are signed by the Internationals, and in view of this fact I would suggest that instead of the Pacific Northwest Association drawing up a program, that the International Union submit the program, if it be for a joint wage conference, let it be submitted.[44]

This recommendation marked the start of a period of some unrest on the West Coast which culminated in a greater role in negotiations for the Internationals, along the lines suggested by John Sherman. In April 1937, Sherman telegraphed Burke:

> Organization formed at Olympia District Council. Agreed to split mills in 3 groups if this is not done withdraw from Association and negotiate alone tentative agreement to hold back per capita tax from International and endeavor to obtain CIO charter.[45]

Burke reacted vigorously, writing Sherman:

> Why any of these locals should be disloyal to the Pacific Northwest P & PMEA or the International Unions is more than I can understand. How about the obligation that these men took when they joined the Unions? They pledged themselves to support the laws and rules and the regulations of the International Unions. Does that not mean anything to them? It would seem to me that there must be a fair percentage of our members at Shelton, Everett, Longview and Hoquiam who have some appreciation and understanding of what has been done by the organization during the past three years. It is hard for me to believe that the majority of the members in these mills favor withholding the per capita tax and joining the CIO, if they cannot get everything they think they ought to have. Does the CIO always get its maximum demands? Did the United Mine Workers get their maximum demands in the settlement just made with the mine owners? What union gets its maximum demands?
>
> As far as withholding the per capita tax is concerned that does not frighten me at all. Our International Union will continue to do business at the same old stand even if we do not get a penny from Shelton, Everett, Longview and Hoquiam.[46]

Evidently Burke's strong reaction had some effect, at least during 1937. At the August meeting of the Employees Association the Executive Board recommended that "the International Officers shall be a part of the negotiating or bargaining board and shall take the lead in negotiating with the employers."[47]

44. Papers, John Sherman to John Burke, February 15, 1936.
45. *Ibid.*, Telegram from John Sherman to John Burke, April 4, 1937.
46. *Ibid.*, John Burke to John Sherman, April 5, 1937.
47. *Ibid.*, "Minutes of the Executive Board Meeting of the Pacific Coast Pulp and Paper Mill Employees Association, August 24–25, 1937."

The next year another change developed in the structure of the Association. The locals of the Paper Makers voted to establish a Pacific Coast Council and the Pulp, Sulphite locals set about establishing district councils. The Executive Board of the Employees Association was worried about the prospect of fragmentation of the organization, and on March 23, 1938, Maxwell Loomis, secretary of the Association wrote President Burke, requesting a charter for the Pacific Coast Pulp and Paper Mill Employees Association.[48] On April 1, Burke wrote Loomis that it was impossible to issue a charter to the Employees Association, that only locals could be chartered. After three and one-half months, Loomis continued the dialogue with Burke:

> I have sometimes wondered whether the Association is not viewed with misgivings in the offices of the International Unions, I suppose there is some danger of dual unionism here in this set-up, but I believe that could we be chartered under the two Internationals, the Internationals could exercise more control than is possible under the present set-up.[49]

Burke again refused to issue a charter to the Association.[50] Evidently this ended the talk of a charter and of the possibility of dual unionism, at least for some sixteen years.

Although a charter for the Association was not granted, schismatic tendencies were manifest in the formation of the Paper Makers Pacific Coast Council and in various district councils of the Pulp Workers. The councils were to be part of the Association and not separate units.

During the late 1930s unrest was evident in other ways. Hoquiam Local 169 and Bellingham Local 309 (IBPM) voted to disaffiliate from the Association. Reaffiliation occurred in 1939, although the records do not indicate when Locals 169 and 309 left the Association. Defections from the Association continued into the 1940s. In 1942, five locals, one from the Paper Makers and four from the Pulp Workers were not affiliated with the Association, and their independent stance moved the Internationals to action.

### *Addition to International Power in 1942*

In early 1942, the International Unions, alarmed at the fragmentation of the bargaining unit developing on the West Coast, were stirred to action. The Pulp Workers Local No. 153 at Longview, Washington, was a source of dissatisfaction with the International. At the negotiations in 1941, President Burke and Vice-President Sherman signed the agreement

48. *Ibid.*, Maxwell Loomis to John Burke, March 23, 1938.
49. *Ibid.*, Maxwell Loomis to John Burke, July 14, 1938.
50. *Ibid.*, John Burke to Maxwell Loomis, July 20, 1938.

on behalf of the International Union, and for the local unions not represented at the negotiations. Among these was Longview No. 153 which had withdrawn from the Association. The action Burke took was clearly in violation of the union constitution.[51] The local asked how it could be prevented from taking part in bargaining when it was a member in good standing of the International. Even though the local was not a member of the Association, it could not be prevented from participating in the bargaining. This situation was one of the factors stimulating the International to seek an increased voice in formulating the agenda for bargaining and in the conduct of negotiations. On January 17, 1942, President Burke wrote John Sherman:

> Vice-President Stephens and I sat in at a session of the Paper Makers Executive Board and we discussed this matter of having the two International Unions call the conference on the Pacific Coast this year. The Paper Makers Executive Board are favorable to this idea.
>
> However, all of us were in agreement that we should not do this without consulting the Executive Board of the Pacific Coast Pulp and Paper Mill Employee Association. We want to get the Executive Board of the Association to see eye-to-eye with us on this. Therefore, I shall want you to attend the meeting of the Executive Board and explain why we think it essential this year to have the conference called by the two International Unions. Vice-President Drummond will be there for the Paper Makers.
>
> A careful reading of the Constitution and By-Laws of the Employees Association has convinced me that unless the conference this year on the Coast is conducted by the two International Unions that we will have to give Longview and any other locals that may desire, separate negotiations if they demand them. We have no power to force local unions into the Association. A local union that is chartered by our International Union has certain charter rights. If we insist upon Longview and these other locals being a part of the Uniform Labor Agreement, then we cannot exclude them from the conference where the program is prepared to present to the Employers. The Uniform Labor Agreement is signed with the two International Unions. Therefore, it seems to me that all the locals that come under the Uniform Labor Agreement have a right to demand that the two International Unions conduct the conference or they have a right to ask for separate negotiations if they want them.
>
> I think that if the two International Unions call the conference that the chances are that all the locals will go along.[52]

At the meeting of the Executive Board of the Association in Portland in February 1942, the board expressed willingness to accommodate the Internationals, in their desire for an increased role in the West Coast bargaining pattern. However, the approval of the executive board was not

51. *Ibid.*, "Minutes" of the Conference of the Pacific Coast Pulp and Paper Mill Employees Association, Portland, Ore., May 20, 1942, p. 11.
52. *Ibid.*, John Burke to John Sherman, January 17, 1942.

granted without misgivings, as the following letter from Maxwell Loomis to John Burke on February 10 demonstrates:

> The Executive Board of this Association has just met in Portland, Oregon, and has given serious consideration to the advisability of the two International Unions calling the pre-conference meeting this year to allow Local Unions, parties to the Uniform Labor Agreement, but not now affiliated with this Association, an opportunity to participate in the preparing and bargaining of agreement changes.
>
> The Executive Board is very doubtful that our Association can survive such a procedure; at the same time, we recognize the paramount importance to the welfare and security of our membership of maintaining the continuity of the Uniform Labor Agreement, and the danger to this continuity inherent in the bargaining of separate agreements for several locals now parties to the Uniform agreement. The following statement was adopted by the Board and reflects the Board's position:
>
>> The Executive Board of the Pacific Coast Pulp and Paper Mill Employee's Association is willing to co-operate with the two International Unions in the following briefly outlined program: The two International Unions will call and conduct the pre-conference meeting at which time changes to the agreement are discussed; also the sessions involved at which time the Uniform Labor Agreement is negotiated.[53]

The grudging acquiesence of the executive board in February did not spell the end of discussion on the matter. At its meeting of March 27 and 28, the Association's board produced a three-page document justifying its position on the matter. The board advanced three reasons for granting power to the International:

1. The paramount importance to the welfare of our membership in preserving the continuity of the Uniform Labor Agreement, especially during the present period of inter-union raiding.
2. Certain National Labor Relations Board rulings which effect [sic] the negotiations of group labor agreements such as ours.
3. The fact that several local unions—one Paper Maker Local and four Pulp Worker Locals—which are at present working under the Uniform Agreement are not now affiliated with the Association.[54]

This document was evidently designed to be circulated among the West Coast locals, and to provide a rationale for ceding power to the Internationals. Prior to the negotiation in May 1942, the Association called a convention for the purpose of discussing the recommendation of its executive board. In recommending acceptance of the board position, Maxwell Loomis concluded:

53. *Ibid.*, Maxwell Loomis to John Burke, February 10, 1942.
54. *Ibid.*, "Minutes" of the Meeting of the Executive Board of the Pacific Coast Pulp and Paper Mill Employees Association, Portland, Ore., March 27–28, 1942, p. 1.

> Now, the Executive Board of the Association did not like the idea of surrendering the Pre-conference Convention to the International Unions, but due to the fact that we are not a legal part of the International set-up, we have no legal standing. So the Executive Board has come to the conclusion, that it is advisable to go along with the program this year as outlined. . . .[55]

There was considerable resistance to the proposal, and Pulp, Sulphite Vice-President Sherman was called upon to outline the position of the International:

> There were four or five locals that were not represented at the pre-conference [in 1941]. Last year it was necessary to make a statement for the signing of the agreements for these particular locals, which signing I undertook for these unions who were not represented at that conference. As a result for some time there was considerable trouble throughout the union. These locals tried to find ways and means whereby they could nullify the contract which was signed by the International Unions. After many months, President Burke took the matter up with the Representatives on the Pacific Coast and he stated very definitely that this year he would not sign any agreement for any local union that did not have the opportunity of sending delegates to the pre-conference here at Portland.[56]

After extended debate, the motion to accept the recommendation of the executive board increasing the responsibility of the Internationals, was accepted 58 to 35.

## *Hoquiam, Washington, Local 169*

The grant to the Internationals of increased power did not reduce local unrest on the West Coast. At the end of 1942, the most serious threat to the Uniform Labor Agreement developed as a result of a situation at Hoquiam, Washington. The pulp workers at Hoquiam were organized into Local 169, IBPSPMW, and employed by Rayonier, Inc. At the end of 1942, the company became involved in a dispute with its employees over its failure to pay overtime rates for work performed on the days following Christmas and New Year's Day. On February 1, 1943, Vice-President Sherman notified President Burke of the unrest at Hoquiam, and concluded:

> We will watch the situation carefully, but I do not believe there is anything to worry about.[57]

Sherman's assessment of the situation soon proved incorrect. On February 25, Sherman telegraphed Burke: "Held meeting Hoquiam Tuesday

55. *Ibid.*, "Minutes" of the Convention of the Pacific Coast Pulp and Paper Mill Employees Association, Portland, Ore., May 20, 1942, p. 4.
56. *Ibid.*
57. *Ibid.*, John Sherman to John Burke, February 1, 1943.

members held secret ballot in mill and voted 171 to 84 to settle [sic] connections with International."[58] Burke responded quickly, and indicated clearly the strategy to be employed by the Pulp Workers. On February 25, he wrote Sherman:

> Fortunately our agreement does not terminate until June 1st. We can require the Company to live up to the terms of the agreement. I think we should be in a strong position to oppose any election being granted by the National Labor Relations Board even though they are signing CIO cards.[59]

The initial vote of Local 169 to disaffiliate from the Pulp Workers made no reference to affiliating with any other union. However, the International Woodworkers of America, CIO, were quick to take advantage of the situation. On March 11, Sherman sought authority to expel 10 members of Local 169 for "promoting and assisting formation of rival unions."[60] On March 12, Burke gave his approval.[61] The same day, the IWA filed a petition with the National Labor Relations Board requesting establishment of a separate bargaining unit composed of Rayonier employees at Hoquiam. On October 9, 1943, the National Labor Relations Board rendered its decision in regard to establishment of a separate bargaining unit for the Hoquiam members of IWA Local 3-362. The Board stated:

> The fact that no active member of the Employer association group has seen fit to bargain independently with the representatives of his own employees demonstrates that the individual employers believe joint collective action by both employers and employees to be the most effective method of bargaining.
>
> The fact that the members of the (employer) association have each year regularly participated in negotiations, with the understanding, pursuant to formal resolution at the pre-wage conference, to abide by a majority decision of the members, clearly shows the assumption of an obligation to adhere to the resultant uniform agreements. The customary observance of such obligation by association members, together with their established practice of otherwise acting jointly in regard to labor relations, and the fact that the employees as well as the companies have long considered collective bargaining on an association-wide basis to be desirable coupled with the other related circumstances hereinabove referred to, impel the conclusion that a unit coextensive with the employees in all association companies is appropriate in the present instance. Accordingly, we find that the single employer unit requested by the CIO is inappropriate for the purposes of collective bargaining.[62]

58. *Ibid.*, Telegram from John Sherman to John Burke, February 25, 1943.
59. *Ibid.*, John Burke to John Sherman, February 26, 1943.
60. *Ibid.*, Telegram from John Sherman to John Burke, March 11, 1943.
61. *Ibid.*, Telegram from John Burke to John Sherman, March 12, 1943.
62. In re Rayonier, Incorporated, Grays Harbor Division (Hoquiam, Washington) and International Woodworkers of America, Local 3–362 (CIO). Case No. R-5754, Oct. 9, 1943, 52 NLRB 1269.

The record cited in this chapter brings into question the wide-spread belief that the collective bargaining relationship in the Northwest paper industry has been extraordinarily peaceful. That harmony existed at the top levels of the International hierarchy and company management is undeniable. However, in the process of achieving a harmonious relationship, certain desirable features may have been reduced to a subordinate role.

In a study of this collective bargaining relationship Clark Kerr and Roger Randall praise the system that evolved on the Coast. But they note that absence of friction is not the sole measure of a healthy labor relations.[63] The quality of the relationship is also important. Kerr and Randall feel that if the workers are inadequately represented, the union dominated by the employer, the employer subjected to unduly burdensome conditions, or purchasers exposed to labor-management collusion to raise prices, the cost of peace may be greater than its value. The costs of harmony may also outweigh the value when the international officers act to stifle the expressed desires of the local membership. The record in this industry seems to indicate that the praise with which this relationship has been viewed was perhaps excessive. The pattern that emerged on the Pacific Coast shows a continuing growth of International power at the expense of local power.

Despite efforts to reassure the locals, the Internationals clearly acted to reduce the sphere of influence of the regional association during the latter half of the 1930s and into the 1940s. International control of the pre-wage conference and the negotiating session marked a great centralization of power. The decision of the National Labor Relations Board in October 1943, had the effect of further increasing the strength of the International at the expense of the locals. Sentiment for disaffiliation was clearly in the majority at Hoquiam, yet the decision of the board prevented a change in representation.

There is a fine line between the power needed for discipline of membership and stable industrial relations, and abuse of such power to thwart the desires of the membership. The record in this situation seems to indicate that perhaps the continuing addition to International power crossed that line, at least in the minds of the rank and file, for in the 1950s rumblings for reform were heard once again.

---

63. Clark Kerr and Roger Randall, *Causes of Industrial Peace Under Collective Bargaining, Crown Zellerbach Corporation and the Pacific Coast Pulp and Paper Industry* (Washington, D.C., 1948), p. 2.

# 3

# FERMENT ON THE WEST COAST

## *Early Years of the Research and Education Department*

At its convention in Toronto in 1944, the International Brotherhood of Pulp, Sulphite and Paper Mill Workers voted to establish a Department of Research and Education "to provide closer contact and instructional material to assist the local unions to function properly."[1] James P. Nicol, Jr., was appointed director for Canada with offices in Montreal, and George W. Brooks was engaged to head the operation in the United States, with offices in Washington, D.C.

The two Research and Education Departments of the Pulp Workers and Paper Makers collaborated and developed joint programs of local training. The programs were integrated into the structure of the unions through the involvement of the International hierarchy. Classes for teacher training of local members were initiated with the request and cooperation of regional offices, and local members were trained as instructors because of the lack of staff and funds on the International level. Training of local instructors insured the development of skilled teachers in the locals, and gave the training program continuity in time. Local instructors were selected by their local union, by whatever means they desired, and the program was financed by the International and the local.[2]

The local unions on the West Coast, acting through their Employees Association, were very interested in training. At the 1947 meeting of their Association, the retiring secretary gave the executive board report recommending sponsorship of an educational program, and the board suggested adoption of a plan to award scholarships in three states. Action was postponed on this suggestion.[3] In 1950, the Association sponsored two union seminars, with classes at the University of Washington. The Association paid tuition fees of thirty-five dollars per person, with the

---

1. International Brotherhood of Pulp, Sulphite and Paper Mill Workers, *Pulp, Sulphite and Paper Mill Workers Journal*, March-April 1945, p. 2.
2. George W. Brooks and Russell Allen, "Union Training Programs of the AFL Paper Unions," in U.S. Department of Labor, Bureau of Labor Statistics, *Monthly Labor Review*, April 1952, pp. 395–399.
3. Pacific Coast Pulp and Paper Mill Employees Association, "Minutes of the Executive Board Meeting, January 28, 1958," p. 11.

locals absorbing the remainder of the cost.[4] The next year, the International Research and Education Departments introduced steward training on the Pacific Coast.

While the training programs were supposedly initiated by the hierarchy, not all the International officers were sympathetic to them, or to the Research and Education Department. In 1953, Pulp Worker Vice-President Ivor Isaacson wrote President Burke:

> I have become so thoroughly disgusted and disappointed with our Research Departments that I will not approve of any additional help or cost for either Brother Nicol or George Brooks. The cost of these two departments has far exceeded what was first proposed. Perhaps not too excessive with Brother Nicol's department, but no question about excessive cost in Brooks' department. As far as I am concerned our Research Department, especially in Washington, D.C. has done very little of value for me or this office.[5]

Even President Burke had some misgivings about the training programs his staff department was conducting. In 1955, he wrote Vice-Presidents William Burnell, John Sherman, Homer Humble, and Louis Lorrain, the members of the union Research and Education Committee:

> I sometimes wonder if we aren't getting in too deep with these training programs. There is a limit to what our International Union can do in providing training classes for our local unions. We cannot keep adding men to our staffs. I am sure that George Brooks means well but he doesn't seem to understand that there is a limit to the amount of money that the International can spend for our Research and Education Department.[6]

## *Unrest in the Employees Association in 1955*

In 1955, the Employees Association, which had given International Headquarters some difficulty in the late 1930s and early 1940s began to increase its activities, both in the area of training and in the fields of collective bargaining and general union policy. John Sherman became alarmed at the activities of the Association, and on October 31 wrote President Burke that the Employees Association:

> . . . has during the last few months been stepping up its activities and the vice-president [sic] in this area have had to watch very carefully so that it does not interfere with the policy laid down by our International Union, particularly that relating to collective bargaining. In the discussions at their meetings they have discussed many important questions relating to bargaining and of course the question of amalgama-

4. *Ibid.*

5. Ivor Isaacson to John Burke, September 8, 1953. Papers of the International Brotherhood of Pulp, Sulphite and Paper Mill Workers, Wisconsin State Historical Society, Madison, Wis.

6. Papers, John Burke to William Burnell, John Sherman, Homer Humble, and Louis Lorrain, August 3, 1955.

> tion [of unions in the industry] has also come before the delegates. We have been successful up to this time affording [sic] any movement which would implicate our International Union.
>
> I hope that we will take this information in the light in which it is given as it is my desire to protect ourselves at this time.[7]

When the Uniform Labor Agreement was renegotiated in 1955 there were 10,451 valid ballots cast. *Yes* votes totaled 4,451 against 3,000 *no* votes. However, certain locals which were later to manifest dissatisfaction with their International voted down the agreement. For example: Local 169 Hoquiam, Washington, voted 206 to 185 for rejection; Local 183, Everett, Washington, 132 *no*, 58 *yes;* Local 713 San Joaquin, California, 155 *no*, 25 *yes;* and Port Angeles, Washington, 186 *no*, 139 *yes*.[8]

## *The 1955 Training Program on the West Coast*

In 1955, the Research and Education Department proposed to conduct a training session on the West Coast.[9] Research Director George Brooks proposed adding a man to the staff to handle the program in the area. He wrote Vice-President Sherman on March 24, "that in my opinion it is a mistake to proceed with any further educational activities on the West Coast unless we can have a man *out there* who will give his time and attention to this activity."[10] Sherman informed President Burke of Brooks's position, but Burke was adamant about not adding a new professional person to the staff.

In July, Brooks wrote Sherman and suggested an alternative to adding additional professional help. His approach was to increase the involvement of the Pacific Coast Pulp and Paper Mill Employees Association.

> Since the Pacific Coast Pulp and Paper Mill Employees Association appears to be very much interested in educational work, we might find a solution through them. Perhaps instead of financing a seminar, the Association would want to compensate someone who would spend his time on educational work, working closely in cooperation with you and with us.
>
> If the International Union cannot undertake the program of education that is needed, we ought to acknowledge this fact frankly.[11]

Vice-President Sherman was strongly opposed to involving the Em-

---

7. *Ibid.*, John Sherman to John Burke, October 31, 1955.
8. *Ibid.*, H. L. Hansen file, 1955.
9. In the 1955 training program a number of people were involved who later became opponents of the International. Local 244 sent Arthur Farrace, Local 713, Clarence W. Dukes, Local 169, Graham Mercer, Local 433, and British Columbia sent Orville Braaten. These men later became active in reform movements within their union, the Pulp, Sulphite Workers.
10. Papers, George Brooks to John Sherman, March 24, 1955.
11. *Ibid.*, George Brooks to John Sherman, July 6, 1955.

ployees Association in the educational program or having the Association compensate anyone for educational work. He wrote to Brooks:

> I am firmly convinced that if such activity were carried on it will destroy the ULA and would take away from the International Union its collective bargaining rights which have been sustained by the Government Board.[12]

Sherman proposed field staff men in Oregon and Washington, British Columbia, and California to follow-up the training program. President Burke approved Sherman's plan, but Brooks was skeptical about the proposal to employ three field men. He agreed, however, to try to work with the system.[13]

The training representatives selected were Stan Green for British Columbia, Oren Parker for California, and John Eyer for Washington and Oregon. The selection of Eyer is perhaps significant. He was to become a supporter of the West Coast reform movement, and later broke with the Pulp Workers to work with the group that disaffiliated in 1964.

It appears that as early as the middle 1950s some dissatisfaction was developing among the hierarchy of the Pulp Workers concerning their Education Department. There was suspicion with regard to supervising the professional staff, and great reluctance to add to that staff. Future years were to witness more concrete fears by the International officers about the growing surge of interest in training displayed by the West Coast locals. Concurrently, the activities of the Employees Association began to cause concern among the leadership. The Association was actively discussing the question of merger of unions in the industry, and was plunging deeper into the field of education and training.

At the same time that there was ferment over education and other matters, complaints began to trickle into International headquarters from the West Coast over job analysis. Pulp, Sulphite Local 236 in Everett, Washington, passed a resolution in February 1955 declaring that it viewed the job analysis system with doubt and was not pleased with its results.[14] Later in the month, the local secretary wrote Burke amplifying the complaints over job analysis: "It seems that job analysis methods and rules as they now exist are not flexible enough to meet the ever-changing conditions that are taking place in the pulp industry."[15] This was an early complaint about job analysis that was to swell to a chorus in the ensuing months and years.

---

12. *Ibid.*, John Sherman to George Brooks, July 25, 1955.
13. *Ibid.*, George Brooks to John Sherman, August 1, 1955.
14. *Ibid.*, In file of Everett Local 236, February 10, 1955.
15. *Ibid.*, Ernest Hedereich, secretary Pulp, Sulphite Local 236 to John Burke, March 5, 1955.

## Developments in 1956

Nineteen-fifty six marked the fiftieth anniversary of the Pulp, Sulphite Workers and was a convention year. It was also the year in which large scale unrest within the union manifested itself. Discussion was lively on a number of issues, from the training programs and job analysis to alleged malpractice in health and welfare programs. In this same year there was also an extremely close vote of acceptance of the Uniform Labor Agreement, and some sentiment within the union for regional election of vice-presidents. The organizing tactics of the International, particularly on the West Coast, also left themselves open to question.

## Training in 1956

A significant divergence of views began to develop in 1956 with respect to the work of the Research and Education Department. The International hierarchy showed evidence of disenchantment with the Department, and indicated a desire to restrict its activities. George Brooks wished to add an additional professional staff member to his department, but Vice-President Sherman was against the idea and told Burke,[16] who agreed that a new professional staff person should not be hired.[17]

In addition to Vice-President Sherman, hostility to the Research and Education Department was manifested by Ivor Isaacson, at that time the seventh vice-president, who worked mainly in California. Isaacson felt that the work of the Research and Education Department was undermining the activity of the vice-presidents and International representatives. On September 5, Isaacson wrote Burke and complained at length about the work of Brooks and the Research and Education Department:

> I have always been opposed to the field manual being sent to anyone outside our staff. The more I see and hear of this sort of bunk, the more convinced I am that I am right in my thinking and I have also come to the conclusion that the sooner we permanently park Mr. Brooks and his staff in their headquarters, the better off we will be. This is the second time that he and his staff have held over all classes on the West Coast and the results are showing up fast. This nonsense of stewards for one and look at the resolutions you are getting from the locals out here. John, I think it is time that the International Executive Board or the Research and Education Committee or someone lay down some laws, rules and regulations for our Research and Education Department to work by. Now let's put a stop to this chasing around the country filling people full of false hopes and belittling the work that we are trying to do. I do not say this in

16. *Ibid.*, John Sherman to John Burke, April 16, 1956.
17. *Ibid.*, John Burke to John Sherman, April 19, 1956.

> criticism of you or the Research Committee's past performance, but I do wish that more force would be put into effect governing this department.[18]

At the same time some members of the hierarchy were becoming disillusioned with the activities of the Research and Education Department, the locals on the West Coast began showing increased interest in the Department. Nineteen fifty-six was a convention year and some of the Western locals submitted resolutions favoring establishment of a Western Division of the Department of Research and Education to serve the locals on the Pacific Coast. Among locals submitting resolutions on the subject were Local 249, Antioch, California; Local 169, Hoquiam, Washington; and Local 153, Longview, Washington.

The resolutions proposing a Western Division of the Research and Education Department stirred the International Officers to opposition. Burke wrote Ivor Isaacson that he was against such a move, as was Isaacson.[19] On August 21, Sherman wrote Brooks questioning the campaign for the Western Research and Education Department:

> I have just received a copy of [a] letter dated August 7, 1956 mailed to all local unions including Paper Makers on the Pacific Coast and requesting each and every local to send a resolution to our International Convention for the purpose of setting up a Western Division of the Research and Education Department.
>
> The letter states that the Executive Board of the [Employees] Association had talked this matter over with the Research and Education Directors of both unions and [I] assume that this was done during the seminar at Portland, Oregon recently. The letter to the local unions goes on to state that it would be impossible for the Western Division to be sponsored by the Association and that the Eastern Division be under the auspices of the two International Unions. It was my opinion and always has been as a member of the Executive Board and also of the Research and Education Committee, that there is no such thing in our International Union as an Eastern Division of Research and Education. It has always been my understanding that the Research Department, both in the U.S. and Canada, services all of our local union [sic] to the best of their ability. I would have to admit however, that both departments have gone far afield from that which originated in 1944. I can assure you that I am deeply disturbed over this matter because the men responsible in this area for the operation of the International Union is [sic] hereby placed in a spot.[20]

Brooks replied that the idea of the Employees Association was unworkable and that there should not be an Eastern and Western Division of the Department. He concluded that, "The International Union set up a department to serve the officers, the staff and local unions throughout the

18. *Ibid.*, Ivor Isaacson to John Burke, September 5, 1956.
19. *Ibid.*, John Burke to Ivor Isaacson, September 11, 1956.
20. *Ibid.*, John Sherman to George Brooks, August 21, 1956.

country and this arrangement is certainly the only one that makes sense."[21]

The Research and Education Committee at the Convention voted not to bring the resolution on a Western Division to the floor. Instead, a substitute resolution was read by Committee Secretary Reynold Victor, who was also secretary of the Employees Association. It provided for the vice-presidents on the Pacific Coast to assign one of their staff members to work full time on research and education under the direction of the Research and Education Department. After perfunctory debate the resolution was adopted.[22]

In its preparation for the convention, the Executive Board deliberated proposing a 25 cent increase in the per capita tax. In the course of their deliberations on the matter Vice-President Ivor Isaacson wrote President Burke:

> The breakdown on the 25 cents per capita is interesting; however, if a stop is not put to George Brooks and his ideas, he will use up the entire 25 cents. This fellow disturbs me. I am afraid that he again has made some commitment out here during these training classes that will cause us no end of trouble. If he keeps on, we will all be working for him.[23]

It seems clear that in 1956 the Research and Education Department was becoming the source of some dissatisfaction to the Executive Board. On the other hand the West Coast locals were very conscious of the work of the department and desirous of increased service. These two divergent tendencies were to have important consequences in the history of the union.

## *Vote on the 1956 Uniform Labor Agreement*

In 1956, West Coast dissatisfaction was manifested in another area. This was the negotiations on the area-wide contract, the Uniform Labor Agreement. After the balloting on the contract was concluded, the final tally showed passage by a vote of 7,340 to 5,200.[24] Sherman did not seem particularly upset by the close vote. He wrote Burke notifying him of the ratification, and Burke was somewhat puzzled by the close result. He wrote that "I am very much surprised at the large no vote. It is hard to understand."[25] It is significant that, as Vice-President Sherman noted,

---

21. *Ibid.*, George Brooks to John Sherman, August 30, 1956.

22. International Brotherhood of Pulp, Sulphite and Paper Mill Workers, *Proceedings of the Twenty-Fourth Convention*. Milwaukee, Wis., September 24–29, 1956, p. 282.

23. Papers, Ivor Isaacson to John Burke, August 30, 1956.

24. The vote was 7,340 yes to 5,200 no. The breakdown was:

| *California* | *Yes* | *No* | *Oregon* | *Yes* | *No* | *Washington* | *Yes* | *No* |
|---|---|---|---|---|---|---|---|---|
| Pulp Sulphite | 812 | 838 | | 984 | 850 | | 3,305 | 2,283 |
| Paper Makers | 917 | 566 | | 346 | 186 | | 976 | 477 |

25. *Ibid.*, John Burke to John Sherman, June 4, 1956.

"many . . . large local unions voted in the negative and evidently were not pleased with the settlement."[26] Some of the locals that rejected the contract were later to become active in urging changes within the International. The margin of rejection in many cases left little doubt as to how the men felt. Local 713, San Joaquin, California, 51 *yes,* 183 *no;* Local 249, Antioch, California, 108 *yes,* 386 *no;* Local 68, Oregon City, Oregon,[27] 461 *yes,* 661 *no;* Local 153, Longview, Washington, 208 *yes,* 460 *no.* Local 580 in Longview accepted the contract by 2 votes out of 510 cast. A number of smaller locals rejected the contract decisively, and the vote in most other locals was very close.[28]

Evidently the delegates to the negotiations did not take kindly to the leadership of their International officers. Ivor Isaacson wrote to Burke, complaining of his experience with the delegates and some other staff members.

> I have attended the Uniform Wage Conference for many years but without doubt this one just completed was by far the most difficult we ever had. The employers were as friendly as usual, not at all abusive. Grimes [Sid Grimes, employer representative] conducted himself in [a] very fine manner but the delegates were just impossible to control. I never saw anything like it. These wage conferences can really wear one down when things go normally, but when you have 126 delegates divided into several factions working against each other and in most instances working against the ones who are trying to make a settlement, well, why tell you—you have been through it. However, a good settlement was finally made. If we could have gotten the delegates to agree to cents per hour increase, as Sherman and I suggested, we would have come out with a base rate of $2.00 or very near to it. However, the majority wanted percent increase and, as you know, that was the request made to the employers.[29]

Burke again did not know what to make of Isaacson's report. He replied to Isaacson:

> I really do not know what to say about your report of the factionalism that was present in the conference. It is hard to understand. I do not suppose that any group of workers have ever made greater material gains than have those covered by the uniform agreement but it is shocking to realize that no gains have been made in the development of a greater spirit of fraternity and . . . brotherhood.[30]

26. Papers, John Sherman to John Burke, June 1, 1956.
27. Local 68, Oregon City, Ore., is the home local of William Perrin who was to lead the disaffiliation move in 1964.
28. Papers, John Sherman to R. L. Gregory, secretary, Local 297, Port Mellon, British Columbia, June 6, 1956.
29. *Ibid.*, Ivor Isaacson to John Burke, May 25, 1956.
30. *Ibid.*, John Burke to Ivor Isaacson, May 28, 1956.

## *Regional Election of Vice-Presidents*

In 1956, the issue of regional election of vice-presidents, which had been raised by the West Coast locals in the 1930s, came to the fore once again. Article III, Section 3 of the Pulp Workers Constitution provides that "the election of International Officers . . . shall be held at the Convention." Section 4 requires that "Officers of this International Union must receive a majority of all votes in order to be elected."[31] Within the union, the practice had developed of having the vice-presidents concentrate their activities in one area or region. Thus, John Sherman, for example, was in charge of operations in the United States Pacific Northwest and Western Canada, including British Columbia, Ivor Isaacson was responsible for California, and William Burnell for Eastern Canada. The areas served by the vice-presidents were ill-defined, and situations developed which allowed those from one area to conduct union business in regions outside of their normal jurisdiction. The practice was particularly true in the earlier years of the organization, when vice-presidents were sent around the country as the need arose. The activities of Herbert Sullivan in the organization of the West Coast are an example of such procedure. In more recent times, perhaps from about 1940 onwards, vice-presidents have tended to remain more in their own areas, and the informal regions have become somewhat more defined, though never set down in writing.

With the union constitution providing that officers were to be elected at convention and had to receive a majority vote of all the delegates, it was possible to envision a situation in which the delegates from the area served by the vice-president would be in favor of one particular candidate, yet receive a different vice-president if the delegates from the rest of the United States and Canada voted against their man. The same situation could arise if members of an area were opposed to a particular candidate. They might be required to work with a vice-president they had voted against, if the remainder of the delegates voted for the man.[32]

In 1956, the issue of regional election of vice-presidents assumed special importance to the locals on the West Coast. In June, International Representative John Eyer wrote President Burke:

> At the conference in Portland, several delegates asked me if I would consider running against [John] Sherman. . . . Let me state that

31. International Brotherhood of Pulp, Sulphite and Paper Mill Workers, *Constitution and Bylaws*, p. 11.

32. This situation occurred in the 1965 Convention when Godfrey Ruddick, a long time vice-president, was opposed for re-election by a candidate selected by the Executive Board. Ruddick, a supporter of internal reform, was defeated for re-election, although an overwhelming majority of the delegates from his area voted for him. This action confirmed the fears of reform supporters.

> when the delegates approached me, I advised them of the practical political considerations involved in unseating any incumbent V. P. This has revived the talk about the regional elections to which I know you are opposed. I respect your position in this and will make no moves to further this matter other than giving factual information to those who might request it.
>
> The above raises two vital questions which disturb me deeply and I would like to pose them as follows: If regional election of Vice-Presidents is wrong on the grounds that it might tend to build little kingdoms, how can we justify the unrestricted power now held by a V. P. in an area. If his sovereignty in the region is to be unchallenged and appeal from his decisions denied, then should not, in fairness, those who are his subjects have a greater voice in his selection?[33]

Burke replied to Eyer, using the same arguments he had used twenty years earlier when the issue was raised by the West Coast locals:

> You mentioned electing Vice-Presidents regionally. I know that this is desired by some of our locals. I have my doubts if electing Vice-Presidents by regions would prove to be as satisfactory as our present system. How could the country be divided into ten regions? For instance, what would be Vice-President Stephens' region? At present he comes from a wide territory in both the United States and Canada. And what would be Vice-President Isaacson's region? Would he be confined to just the one state of California? I raise these questions just to show that it it not an easy matter to set up regions. You asked the question if regional election of Vice-Presidents is wrong on the grounds that it might tend to build little kingdoms. How can we justify the unrestricted power by a Vice-President in an area? No Vice-President has unrestricted power in an area. No Vice-President has sovereignty over any particular area. The Constitution of our International Union does not give a Vice-President any such power. The members of this organization are not the subjects of any Vice-President or International Officer. It is always possible to appeal to the International President-Secretary [the] decisions of a Vice-President.[34]

Burke's letter to Eyer did not mollify the locals in the West. At the 1956 Convention, resolutions on regional election of vice-presidents were introduced by Local 169, Hoquiam, Washington; Local 708, Watson Island, Prince Rupert, British Columbia; and Local 713, San Joaquin, Antioch, California.[35] After a short debate the resolutions were defeated,[36] but the issue did not die. As time passed, regional election of vice-presidents became a major goal of the reform movement within the International. President Burke was to be bombarded with letters urging its adoption, and when disaffiliation occurred on the West Coast in 1964, regional election was provided for in the constitution of the new organization.

---

33. Papers, John Eyer to John Burke, June 20, 1956.
34. *Ibid.*, John Burke to John Eyer, June 30, 1956.
35. IBPSPMW, *Proceedings of the Twenty-Fourth Convention*, p. 137.
36. *Ibid.*, p. 138.

## *Constitutional Change*

At the convention in 1956, a resolution was introduced by President Burke and Treasurer Frank Barnes to amend the International Constitution. The resolution favored deleting Sections 2, 3, and 4 of Article XVI. Article XVI was concerned with local union representation at wage conferences. Section 3 provided that the basis for local representation at the conferences would be the same as for the convention. Section 4 gave the wage conference, with advice of the Executive Board, "full power to formulate a policy each year to be followed in meeting the manufacturers in conferences and drafting wage schedules and agreements."[37] William Perrin of Oregon City Local 68 rose to speak against the resolution, noting it was going to damage collective bargaining under the Uniform Labor Agreement by ignoring the proportional representation system established for conventions, which Perrin felt represented the membership of the locals.[38] President Burke reassured Perrin and other objectors that local unions would develop their own basis for representation. He concluded his remarks before the vote:

> But it is still a local problem, and whether you decide it or work out the basis of representation at a meeting of your employees association, or at the pre-conference meeting of delegates, you still have to do it there.[39]

With this reassurance, the resolution passed.[40]

## *Local 679*

The formation of Pulp, Sulphite Local No. 679 in New York City in 1956 was to give rise to charges of corruption several years later. In view of the charges of malpractice later levied against Vice-President Joseph Tonelli, the beginnings of Local 679 were rather innocuous. On December 4, 1956, Tonelli sent President Burke a copy of the bylaws of Local 679.[41] On December 11, Burke wrote Tonelli:

> I have received the by-laws of Local 679 and have approved them and sent them to our friend, Tony Barbaccia. Tony certainly has drafted some long by-laws.[42]

A year later Local 679 was discussed by the Association of Catholic Trade Unionists before the McClellan Committee. An exposé was written

37. *Ibid.*, p. 64.
38. *Ibid.*, p. 65.
39. *Ibid.*, p. 68.
40. *Ibid.*
41. Papers, Joseph Tonelli to John Burke, December 4, 1956.
42. *Ibid.*, John Burke to Joseph Tonelli, December 11, 1956.

in the *New York Post* and the entire situation furnished the reform movement with material in its campaign to change the International. It was alleged that the officers of the local exploited the membership, mainly Negroes and Puerto Ricans. Charges were made that the officers were using the union for personal gain, and until the situation was resolved, Local 679 caused the International great difficulty.

## *West Coast Organizing*

The organizing efforts of the International in the 1950s seem to indicate that the union had decided to operate through management in its efforts to recruit new members. Rather than attempting to appeal to the men in the mills, the union leadership dealt with its counterparts in the corporations. The philosophy seems to have been that it would be difficult to attempt to persuade the employees to join voluntarily. Rather, management help was enlisted to pressure people into joining and to prevent competition from other unions. When a new plant is recruiting a labor force, the unorganized employees typically have a voice in determining which union shall represent them. In some instances on the West Coast in the 1950s, the choice of the bargaining representative was made by management acting with the Pulp Workers Union. The opportunity for choice was not presented.

## *Crown-Zellerbach at Antioch, California*

In 1956, the Crown-Zellerbach Company was completing work on a new, three-mill complex at Antioch, California. In March, International Representative Ray Bradford wrote President Burke:

> On April 13 a meeting has been scheduled with top Crown officials and their attorney for a discussion on the two plants at Antioch. The purpose of this meeting is to be sure our strategy is correct and to be positive we are within the law.[43]

In subsequent letters, Bradford warned that other unions were becoming active in the Antioch area.[44] Burke placed his faith in the Crown-Zellerbach management to insure a result favorable to the Pulp Workers. He wrote Bradford:

> I hope the officials of the Crown Zellerbach will be realistic enough to recognize that in this situation the thing to do is to sign up without the formality of NLRB election.[45]

43. *Ibid.*, Ray Bradford to John Burke, March 31, 1956.
44. *Ibid.*, Ray Bradford to John Burke, July 20, 1956.
45. *Ibid.*, John Burke to Ray Bradford, August 1, 1956.

Burke's hope that Crown-Zellerbach would cooperate was realized, for on September 2 Bradford wrote:

> The organizing campaign at Antioch is going along very nicely at the present time. Crown Zellerbach officials are very anxious for us to take these plants and are working very close with me. In fact, no one is hired until I pass on him or her.[46]

On September 6, a contract was signed putting Pulp, Sulphite Local 850 at the Crown mills at Antioch under the Uniform Labor Agreement. Ten days later, Bradford wrote Burke: "I am still passing on all employees hired, in fact I am a behind the scene personnel man."[47]

It is somewhat ironic that in spite of the approval of the Pulp Workers staff for new employees and local union members, Mount Diabalo Local 850 was to become one of the most ardent champions of reform within the International.

Similar organizing "campaigns" were undertaken at a new Diamond Match plant at Red Bluff, California, and at the plants of the International Paper Company's new operations at Turlock and San Jose. Burke had confidence in his tactics, writing International Representative Carol Howes: "I am sure that the management of the International Paper Company will give no trouble in organizing the workers at Turlock and San Jose."[48] He was right. Of the situation at Red Bluff, Ivor Isaacson wrote Burke:

> I met with Mr. McBreen and Mr. Ramsay of the Diamond Match Company relative to their pulp mold plant being built at Red Bluff, California. Mr. Ramsay has been to see you at headquarters on occasion. You may remember him. These gentlemen inform me that it is a must to have our union in this new plant. In fact, they went so far as to say they would fight with all their power to keep out any other union.[49]

The campaigns at Diamond Match and International Paper were concluded successfully in 1957.

The organizing tactics employed by the Pulp Workers seem to indicate that the union and the firms with whom it dealt had a very close relationship. The reasons for this closeness probably are to be found in the fact that the union had been accepted by the companies without a great struggle. It may be that as the relationship continued the union became too disposed to see the companies' point of view. In exchange, the companies supported the unions in their efforts to organize their employees. The results of this harmonious relationship may have produced for the

46. *Ibid.*, Ray Bradford to John Burke, September 2, 1956.
47. *Ibid.*, Ray Bradford to John Burke, September 16, 1956.
48. *Ibid.*, John Burke to Carol Howes, October 29, 1956.
49. *Ibid.*, Ivor Isaacson to John Burke, August 30, 1956.

workers smaller economic gains than if a more "at arms-length position" had been adopted by their bargaining agent.

## *Developments in 1957*

The unrest that was manifest on the West Coast in 1956 continued into 1957. The negotiations on the Uniform Labor Agreement proved very difficult for the Internationals. Vice-President Isaacson was moved to write President Burke:

> Our wage conference in Portland, in my opinion, was the worst of all. Of course, the employers were in a bad mood for negotiations. Their 'that's it' attitude did not help matters at all. I am afraid we will get a strong 'no' vote on the results. If so, we are in for a long drawn out mess. . . .[50]

A few weeks later Isaacson wrote Burke:

> I have just returned from Portland, Oregon where I attended the second Uniform Labor Agreement Wage Conference for this year. As you probably know, it was not at all successful. At least I expected the employers to move to a better settlement. But they did not seem too concerned. Rather a smug atmosphere on their part, I thought. If they would have given us another one-half percent in the increase with the three weeks vacation after ten years for this year, they would have bought a lot of friendly and harmonious feelings without much cost to them. But they did not do this. I think that we will still get a large 'no' vote on this.[51]

Isaacson's fears were borne out, when the final vote on the contract showed 8,658 *yes* votes and 5,469 *no* votes. Burke had little sympathy with the workers' desire for an improved contract. He wrote Isaacson:

> I think the majority acted sensibly in voting for acceptance in preference to closing down the mills. Suppose the workers had rejected the offer of the employers and closed the mills and threw picket lines around the plants. I wonder how much public sympathy they would have, after the employers made known the wage rates with a $2.01 minimum.[52]

Burke appears to have been out of touch with the situation on the West Coast. Evidently the people on the scene felt they had problems that could not be resolved by increasing wage rates. Burke betrayed his remoteness from the situation in a letter to International Representative Cash Price:

> As soon as negotiations are completed, I am going to suggest to Vice-Presidents Sherman and Isaacson that a study should be made of the situation in our local unions up and down the coast to see if we can

---

50. *Ibid.*, Ivor Isaacson to John Burke, May 23, 1957.
51. *Ibid.*, Ivor Isaacson to John Burke, June 13, 1957.
52. *Ibid.*, John Burke to Ivor Isaacson, June 25, 1957.

find out why there is this seemingly [sic] dissatisfaction. It seems strange to me that there should be dissatisfaction among workers who have made so much progress in obtaining so many of the good things of life through their membership in our union and as a result of the negotiations with the employers. How can workers who are able to take one or two or three weeks and in some cases four weeks away from work and get paid for it, harbor these feelings and resentments against their union or someone connected with it? I should think that they would count their blessings and consider themselves very fortunate that there is a union like ours that they can be members of.[53]

## *Dissatisfaction with Job Analysis*

One of the items with which the West Coast locals were concerned was job analysis. The locals had reservations about the system, but Burke favored the plan. He wrote Robert Hetherington, who was the union representative in the system of joint job analysis:

> I regret very much to learn of this unfair criticism from some of the delegates. Our job analysis program had put hundreds of thousands of dollars into the pay envelopes of the workers in these mills. As you say, it is a service that is not given in other areas or by other unions. It makes me wonder what we can do to satisfy some of our members. I doubt if there is any other union in this country that approaches ours in the services given to local unions. When we look back to the year 1934, when the first agreement was signed, and compare wages and working conditions then, with what they are now, it makes you wonder why any members could have anything but words of praise for his union and for those who have represented his union.[54]

In 1957, the union newspaper, the *Paper Worker,* which was the province of the Research and Education Department, ran an article headlined "West Coast Local Eliminates Job Analysis Program." The story concerned the Marathon Corporation plan of job analysis which was administered solely by the company. The union job analyst, Robert Hetherington wrote Brooks indignantly:

> You would be surprised to know that this has caught the attention of many of the people who would destroy the job analysis program under the Uniform Labor Agreement. I feel you should inform your readers that the program of Marathon is strictly a management program confined to their own company and does not resemble in principle or administration the program of joint job analysis under the Uniform Labor Agreement. Your article furnished our critics with ammunition to throw around. . . . Ever since your department circulated that off-base booklet, 'Whats Wrong With Job Evaluation', you could have outlined the difference between company administered plans and the joint program under the Uniform Labor Agreement . . . .[55]

53. *Ibid.,* John Burke to Cash Price, July 6, 1957.
54. *Ibid.,* John Burke to Robert Hetherington, May 3, 1956.
55. *Ibid.,* Robert Hetherington to George Brooks, December 23, 1957. There is no reply by Brooks in the files.

## Training in 1957

The convention in 1956 passed a resolution calling for one staff man to devote his full time to research and education on the West Coast. John Eyer was appointed to this position on a trial basis and proved successful. Sherman advised making the appointment permanent and Burke agreed.

A complaint about the Research and Education Department in 1957 came from the Midwest, and Vice-President Ray Richards wrote Burke on October 29:

> You know, I am wondering how far this training is to be carried out, and oh boy, the subjects. At one time we talked about steward training. That is small stuff today. Last night I received a call from [International Representative] Carl Gear. He told me he had just received a telegram from Brooks asking that he assist as one of the instructors at the training classes I have mentioned. He wanted to know if I wanted him to go. This really burned me up. Who does Brooks think he is, calling my men, especially Gear, to help him in his training program. I was surprised, indeed, believe me. I wish Gear would use some of this knowledge that he possesses in his line of work.[56]

Burke replied that he was in agreement with Richards.[57]

## Genesis of the 1959 Unfair Labor Practice Charge

In early 1957, the Pulp, Sulphite Research and Education Department began to collect material for bargaining with the Manufacturers Association on pension plans. In March, Director Brooks notified Pulp Worker Vice-Presidents Sherman and Isaacson and UPP Vice-President Al Brown that the material submitted by the Manufacturers Association was inadequate for bargaining.[58] A similar letter was mailed on April 16.

It seems evident, from the fragmentary evidence available, that the Manufacturers were reluctant to discuss pensions as early as 1957. Two years later, the International Union would file an unfair labor practice charge seeking to compel the employers to bargain on this issue. The circumstances surrounding the filing of the charge, and the vigor with which it was, or was not, prosecuted were to produce a great deal of discussion within the union.

## Local 679 and Anthony Barbaccia

At a hearing before the Senate Select Committee on Improper Activities in the Labor or Management Field in 1957, John McNiff of the Associa-

56. *Ibid.*, Ray Richards to John Burke, October 29, 1957.
57. *Ibid.*, John Burke to Ray Richards, November 7, 1957.
58. *Ibid.*, George Brooks to John Sherman, Ivor Isaacson, and Al Brown, March 22, 1957.

tion of Catholic Trade Unionists listed Pulp, Sulphite Local 679 as one which cooperated with business to exploit Negro and Puerto Rican workers.[59] When Local 679 was named by McNiff, it produced consternation in the hierarchy of the Pulp Workers. John Sherman commented:

> Certainly, right or wrong, this is not good publicity for our International Union and if the report is not true, it should be corrected immediately. If it is true, then we have a job to do in cleaning up our local unions which are engaged in such activities. It is disturbing, to say the least, to have our union involved in such tactics.[60]

President Burke was outraged that the organization he had given his life to was accused in public of collusion to exploit minority groups. He wrote Anthony Barbaccia:

> I do not know if you realize Brother Barbaccia what a serious matter this is so far as our International Union is concerned. There are several unions in the AFL-CIO that are fighting our union all of the time. They are opposing us in NLRB elections. They are saying all kinds of things against our union. I dare say that the representatives of these unions have cut from the papers this list of local unions that McNiff charged with exploiting the Puerto Ricans and from now on at the NLRB elections they will charge that our union is a racketeering one. That is bad publicity in the papers. That bad publicity in the papers against Local 679 may be [sic] of defeating us in some NLRB elections. This must be cleaned up. Members of our Executive Board are very much concerned with this matter. Unless this can be cleaned up, I will have to have this shop where it is alleged that Local 679 has been exploiting the Puerto Rican workers transferred to some other local. Again, I want to impress upon you the seriousness of this matter.[61]

Barbaccia defended himself by calling his attackers communists.

The episode was to return to haunt the International in 1959 and 1960.

---

59. U.S. Congress, Senate, Select Committee on Improper Activities in the Labor or Management Field: *Hearings, Part 10*, 85th Cong., 2nd sess., 1958, pp. 3756–3781.
60. Papers, John Sherman to John Burke, August 19, 1957.
61. *Ibid.*, John Burke to Anthony Barbaccia, August 21, 1957.

4

# RISING DISCONTENT

## *Negotiations of 1959*

The negotiations of 1959 for renewal of the Uniform Labor Agreement provided the locals on the West Coast with a substantive issue around which to rally their forces. The issue was a request by the unions that the Manufacturers Association bargain on pensions on a unit-wide basis.

At the meetings on the Uniform Labor Agreement in 1958, the union delegates directed the Research and Education Departments of the Pulp Workers and the UPP to make a study of the pension situation under the ULA and report back to the prewage conference. At the meeting in May 1959, a report was made to the delegates, and a pension committee was named to propose a position for the negotiations. The committee report recommended the addition of a new section to the Uniform Labor Agreement which would cover the subject of pensions. The section proposed by the unions, number fourteen on their agenda, favored establishment of pensions for employees of firms party to the ULA. The unions clearly expressed their intention to bargain over pensions with their proposal that:

> No pension plan now in effect or to be established shall be modified or terminated as to the employees in the bargaining unit except through collective bargaining.[1]

The proposal of the unions was submitted to the Manufacturers Association on May 15, 1959. The next day, the manufacturers replied to the unions on the pension question. The manufacturers flatly rejected the union proposal. Their bargaining representative, Sid Grimes, commented:

> Certainly the Union officers know, and presumably the delegates also know, or should know, that this Manufacturers' Association is not the bargaining agent for retirement plans . . . for these companies nor these member mills. Obviously, because of that, your item number fourteen can not be bargained for at this Conference between the Association and the Unions unless the companies which own the mem-

1. International Brotherhood of Pulp, Sulphite and Paper Mill Workers and United Papermakers and Paperworkers, and the Pacific Coast Association of Pulp and Paper Manufacturers, *Record of Negotiations*, May 15–29, 1959, Portland, Ore., p. 28.

> ber mills here and now would abandon their long-established position as to pension bargaining at the company level, and in place thereof would authorize this Association as their legal bargaining agent for pensions.
>
> The Manufacturers are unwilling to make this change. I will briefly review this item, the circumstances under which the authority and the responsibility for bargaining other collective bargaining matters with the full knowledge and consent of the Unions. Sometime in 1944 two of the companies owning a number of our member mills decided they were ready to put a company retirement plan into effect which would include among many others the employees under the Uniform Labor Agreement.
>
> In order to make adoption of those retirement plans possible by those two companies, it was necessary to definitely separate pension plan bargaining from the other bargaining functions of this Association. As I said before, that was done with the full knowledge and approval of the Unions.[2]

In the face of the manufacturers' reluctance to bargain on the subject of pensions at the Association level, the unions called upon Pulp Workers Research Director George Brooks to reply. The research director had general responsibility for preparing the union proposal, but for the first time was taking part in the negotiations. His approach to negotiations appears to have been foreign to both the union and management representatives. Brooks relied on economic and legal arguments to make his negotiating points. At prior negotiating sessions, the discussions had been led by men who had known one another since the early days of organization on the West Coast. Their approach was that of one gentleman to another, with little utilization of expertise. The hard-hitting presentation of the research director may have antagonized the management representatives, but it was evidently what the delegates from the locals wanted to hear.

Director Brooks responded to the manufacturers' refusal to bargain as follows:

> We want you to bargain now. We think that there are millions of dollars of our money, some deposited through payroll deduction and some without payroll deduction, in these funds and we have a vital concern in what happens to it. And our concern goes to the whole range of pension administration and benefits. We are not only concerned with minimum benefits, with benefit formulas. We're also interested in knowing why the rate of return on some plans is so much lower than others. Why one company can pay more than another company although the second company has payroll deductions and the first does not.
>
> All of these things are vital concerns of the employees in these plants. Now if I understand the employers' position here, it is that the

2. *Ibid.*, p. 50.

> Manufacturers' Association has the right to decide unilaterally what things it will bargain about as an Association and what things it will not bargain about. It seems to be asserting that it can bargain about vacations as an association but reserve the right to bargain about holidays to the individual companies. To bargain collectively as a group of companies about Sunday overtime, but on a company basis about Saturday overtime. To give the Association the right to bargain about health and welfare plans but to withhold that right on pension plans. This position is absurd.
>
> The Association is the bargaining agent. It cannot reserve the right to the individual companies to bargain or not bargain about any bargainable matter. We therefore are asking you again to bargain here and now. But before giving us your final answer—which we would like to have as soon as possible—we want to indicate what we regard as the alternative.
>
> We do not propose to take economic action to establish our right to bargain about pensions. It would be foolish to do so, to get something to which we think we are already entitled by law. But if you now say to us "no, we continue in our refusal to bargain about pensions," we shall file an unfair labor practice with the National Labor Relations Board charging this Association and each of its companies with a violation of Section 8-A (5) which requires employers to bargain collectively with representatives of their employees.[3]

This indication of impending legal action if the Manufacturers Association did not change its position on the pension issue was foreign to the spirit that had prevailed formerly at negotiations. Despite the differences that had developed between employer and employee groups over the years, the threat of legal action had never been raised. Now, an outsider, a technician, had stated that an unfair labor practice charge was to be filed. The prospect was unnerving to both sides. The employers were unsure of the legality of their position, and it seems the unions were reluctant to take such a drastic step. However, when the union delegates caucused, the motion was moved and seconded that:

> we [the union delegates] give authority to our Bargaining Board to proceed on our present position on the matter of our pension proposition, item 14, and if a satisfactory answer is not given to the Unions by the Manufacturers' that we file unfair labor practice charges.[4]

Both International Unions stated a charge would be filed. The motion was carried.

The action of the Research Director George Brooks in stating that a charge was to be filed was evidently the wish of a majority of delegates. The position of the hierarchy of the unions is less clear since they never expressed themselves at that time.[5] It seems that the International officers

---

3. *Ibid.*, pp. 65–66.
4. "Minutes of the Union Caucus," May 18, 1959, p. 35.
5. Paul Phillips to John Burke, July 21, 1959. Letter in possession of this author.

were not in control of the situation. The leadership function had been assumed by the visiting research director. Brooks played an active role in the bargaining of most other issues in 1959. He repeatedly led the discussion and challenged the manufacturers' spokesman with an approach that relied on a knowledge of economics unknown in prior negotiations. In a discussion of the proposal on wages, he complimented the manufacturers' representative on stating "so many absurdities with a straight face."[6] He also found some of the manufacturers' suggestions "obnoxious."[7]

With the research director playing a central role in the negotiations, the manufacturers apparently decided they had to act to reduce his importance and to bolster the stature of the International officers. They did this by attacking his statements on the grounds that they were offensive and showed a fundamental ignorance of the operation of the ULA. The manufacturers' spokesman stated:

> You came into this Conference . . . and used such terms that what we said was offensive to you, and what we said was obnoxious to you and what we did was capricious, and what we did was illegitimate to you as a statistician, or something to that effect.
>
> I just want to say George that we haven't reached this relationship we have here over the last twenty-five years by those kind of terms, by accusing each other of misrepresenting things, using deception and the like. I realize that maybe you haven't been out on the Pacific Coast in the Pulp and Paper Industry before on our negotiations. I know a little bit about the negotiations in the East between Labor and Management in the Pulp and Paper Industry, and from that standpoint I am not blaming you. But I am just trying to say that we have not gotten to the position where we are on the Pacific Coast by accusing each other of deceptions, misrepresentations, of being offensive to each other, and obnoxious to each other and such as that.[8]

In addition to the pension issue and wages, local discontent was reflected in other items on the union agenda. Reflecting the fact that the ULA had come to include converting operations as well as primary pulp and paper mills, the delegates proposed adoption of a converting supplement. The supplement was rejected. The unions also requested an increased role for the shop steward in grievances. The delegates voted 67 to 56 to recommend nonacceptance of the contract back to their local unions.[9]

In the conduct of the negotiations in 1959 the Pulp, Sulphite research director played a large role. He had been on the West Coast in earlier years conducting training programs and was evidently well liked by the locals in the area. It seems reasonable to assume that the local union

6. *Negotiations,* p. 168.
7. *Ibid.,* p. 169.
8. *Ibid.,* p. 264.
9. "Minutes of the Union Caucus," p. 10.

representatives were, for the most part, delighted with his vigorous conduct of negotiations, in contrast to the more restrained approach of the International hierarchy. The research director, apparently, took on added stature in the eyes of the delegates. In addition, the stress Brooks placed on local union autonomy and democracy fit in well with the developing dissatisfaction on the local level evident on the West Coast. His actions and philosophy evidently operated to help the budding rebellious elements coalesce.

### *International Reaction to Unfair Labor Practice Charge*

The handling of the proposed unfair practice charge against the Manufacturers Association seemed to confirm the feeling of some locals that their unions had a "peace at any price" philosophy. The president of the Pulp Workers wrote to locals covered by the ULA:

> The members of the Executive Board think it is our duty to give all possible support to our local unions and the local unions of the United Papermakers and Paperworkers in negotiating suitable pension plans, and improving the pension plans of the companies where they are not satisfactory. Therefore, we shall proceed to arrange conferences with the managements of these plants for the purpose of negotiating more satisfactory pension plans.
>
> We will use all honorable means during this contract year to induce the managements of these companies to negotiate better pension plans for our members.
>
> Should the companies fail in their obligations to negotiate improved pension plans, we will give our local unions covered by the Uniform Labor Agreement our full support in making the pension question the chief item of collective bargaining in next year's negotiations.[10]

Burke also wrote UPP President Paul Phillips that the Pulp Workers were hesitant about filing the unfair practice charge because of the effect it might have on relations with the Manufacturers Association and the long duration of the proceedings.[11] Phillips, however, felt that since the commitment had been made to file the charge the unions had to proceed. Phillips also noted:

> The flat statement to the Manufacturers that charges were going to be filed against them should never have been made.[12]

The local unions concerned with the unfair practice charge were upset by the delaying tactics of the Internationals. In August, Burke received letters from Local 153, Longview, Washington, noting the charge had been hailed as a blessed event by the membership. The letters went on

10. John Burke to Paul Phillips, July 17, 1959. Letter in possession of this author.
11. *Ibid.*
12. Paul Phillips to John Burke, July 21, 1959.

to complain about the International officers' alleged milk-toast position and lack of backbone. Local 153 concluded on a note of defiance:

> If they [the International Officers] are more concerned with the friendship of management than the respect of the rank and file, then perhaps it is time we looked for different officers, or another union. There are many of our members who would accept the latter without hesitation.[13]

Letters were received from other West Coast locals accusing the International of sitting on the case and getting soft.[14] One local complained of "prayer meetings" instead of action on the charge.[15]

The correspondence between the locals and International Headquarters on this matter seems to indicate the locals were in no mood to give in to International pressure. Delay in filing the charge was taken as an affront by the locals, and they would not be placated. The correspondence reflects almost open defiance of the International officers. The charge was ultimately filed by the unions, and they were sustained at every level. The Manufacturers Association fought the case to the Court of Appeals before bowing and agreeing to discuss pensions on the Association level. The National Labor Relations Board sustained the unions on October 3, 1961, as did the Ninth Circuit Court of Appeals on June 27, 1962.

## *The Melton Letter*

Dissatisfaction with the settlement in 1959 was manifested in a letter from Secretary Melvin Melton of Bellingham, Washington, Local 194 to the Pulp, Sulphite Workers *Journal.* Printed in full by President Burke, the letter criticized the settlement and made specific suggestions for reform. Secretary Melton felt the unions made a poor settlement in 1959.

> We made a poor settlement because we are victims of a peace at any price contract that is a white elephant. Too many want to maintain this sacred cow, to be able to point to the Uniform Labor Agreement and shout to the world that "we ain't had a strike or one day's labor stoppage in these mills for twenty-five years."
>
> We've paid one hell of a price for these twenty-five years of peace. Maybe it was worth it, if one can afford it.[16]

Melton went on to say the negotiations had become too big and unwieldy and made suggestions for improvement. He advocated regional election of vice-presidents and a voice in selecting field men as well as

13. Fred Delaney, vice-president Local 153 to John Burke, August 18, 1959. Letter in possession of this author.
14. Clarence Dukes, recording secretary, Local 713 to John Burke, August 21, 1959. Letter in possession of this author.
15. Ralph E. Davison, president, Local 161, to John Burke, September 29, 1959. Letter in possession of this author.
16. Pulp Sulphite Workers, *Journal*, July-August 1959, pp. 5–6.

a recall procedure. Melton also wanted the wage conferences removed from the "lair of the employer," Portland. Other suggested changes were a rule that no union representative would discuss matters affecting a local with the employer without a representative of the local present. Melton also advocated changes in the procedure for negotiating the ULA. He wanted delegates to elect a negotiating committee of one delegate from each local with power to accept, reject, or recommend a course of action. Other delegates could go home. The referendum on the contract would be eliminated when the new system was in full effect.[17]

## *The Convention of 1959*

At the twenty-fifth convention of the Pulp Workers, held in Montreal, Quebec, in August and September of 1959, the proponents of change came prepared for battle. Opposition to the hierarchy was rampant, and one noteable victory was gained. As in 1956, a central point raised by critics of the International was the issue of regional election of vice-presidents. Seven resolutions on the subject were introduced at the 1959 Convention. Resolution No. 13 was introduced by Bellingham, Washington, Local 194—Melvin Melton's home local. Melton's resolution was quite lengthy and detailed. He proposed that the United States and Canada be divided into ten areas, and set out the boundaries of each of them, including the states and provinces to be included in each region. Central to the Melton proposal was a suggestion for recall of vice-presidents "on charges of corruption, lack of ability, conflict of interest, and immoral action or utterances."[18] The Constitution Committee recommended nonconcurrence. There was a lengthy debate on the resolution in which critics of the International, such as William Perrin, Oregon City Local 68, Orville Braaten, Vancouver Converters Local 433, and Angus Macphee, Watson Island Local 708, spoke for the resolution which was defeated. Melton had spoken the truth when he said, "I knew damnable well that this resolution would not get anywhere."[19] The other resolutions on this subject were also defeated.

Another item supported by the reform element was compulsory retirement of executive board members at age 65. Three resolutions were introduced on this subject. The Constitution Committee recommended nonconcurrence and the resolutions were defeated.[20]

At the 1959 Convention, for the first time in many years, there was

---

17. *Ibid.*, p. 6.
18. International Brotherhood of Pulp, Sulphite and Paper Mill Workers, *Report of the Proceedings of the Twenty-Fifth Convention*, August 31–September 4, 1959, p. 196.
19. *Ibid.*, p. 197.
20. *Ibid.*, p. 122.

a challenge to an incumbent officer's bid for re-election. First Vice-President William Burnell faced opposition from International Representative Patrick Connolly, whose campaign was managed by International Representative Paul Hayes. Neither Connolly nor Hayes had heretofore played a role in the developing reform movement and Connolly's candidacy evidently surprised some of the delegates who were advocates of change. About halfway through the role call, Connolly conceded defeat. In 1960, both Connolly and Hayes were fired from the International staff.

The Western locals won one noteable victory at the convention, the addition of an eleventh vice-president to represent the Western Canadian Provinces and the State of Alaska. The Constitution Committee recommended nonconcurrence with the resolutions favoring the additional vice-president.[21] The eloquent statement of Delegate Orville Braaten from Vancouver Local 433 prompted President Burke to state he favored referring the resolution back to committee for reconsideration and another recommendation.[22] The committee recommended the executive board study the problem.[23]

Delegates Braaten and Angus Macphee spoke against the revised recommendation and President Burke let it be known that he favored addition of an eleventh vice-president to the executive board.[24] Burke refined his position further. He wanted to let the executive board, at the close of the convention, discuss the matter, and select the new vice-president if it was decided a new position was necessary.[25] This was not agreeable to the delegates. The officers had clearly lost control of the convention on this issue. When President Burke wished to make another statement on the subject he was shouted down by cries of "no."[26] Burke was pushed by the delegates into adopting the position he sought to avoid. The committee report was defeated and a motion adding the additional vice-president was adopted.

There was one other issue that agitated some of the reform delegates at the convention. This was a rumor that Frank Barnes would stand for re-election as treasurer, after which he would resign, allowing the executive board to appoint its candidate to the position. Barnes took to the convention floor to deny the rumor. He said, "if you re-elect me this time, no one is going to appoint Henry Segal as my successor unless God prevents me from staying on the job. . . ."[27] Barnes was re-elected, but in July 1960, he submitted his resignation, effective in October, for reasons

21. *Ibid.*, p. 209.
22. *Ibid.*
23. *Ibid.*, p. 225.
24. *Ibid.*, pp. 225–227.
25. *Ibid.*, pp. 228–229.
26. *Ibid.*, p. 229.
27. *Ibid.*, p. 130.

of health.[28] In an election conducted by the executive board to fill the post of treasurer, International Auditor Henry Segal received eight votes and International Representative Charles Stewart five votes.[29] The fears of the reform delegates were confirmed, although it is impossible to state that Barnes did not resign because of ill health or that Henry Segal's election was fraudulent in any manner. Doubt, however, existed in the minds of some International members.

The events of 1959 set the stage for the increased unrest on the West Coast and throughout the International during the next year. The West Coast locals began thinking seriously about a reform program and developing support for it around the country.

---

28. *Pulp, Sulphite and Paper Mill Workers Journal*, July-August 1960, p. 1.
29. *Ibid.*, p. 2.

# 5

# CHALLENGE TO INTERNATIONAL CONTROL

## *The Dowry*

In 1959, the *New York Post* carried a column by Murray Kempton entitled "The Dowry." The story alleged that a vice-president of the Pulp Workers, Joe Tonelli, had "given" Pulp, Sulphite Local 679 in New York City to the brother-in-law of a Teamster Union official as a wedding present.[1] The name of the brother-in-law was Anthony Barbaccia of McClellan Committee fame. The article alleged that Barbaccia had engaged in improprieties during his tenure as president of the local, and had systematically exploited the membership, consisting mainly of Negroes and Puerto Ricans. In October of the same year, Barbaccia was forced to resign as president of Local 679 at the behest of the International Executive Board. As a reward for his services, he was allegedly granted a pension of one hundred dollars per week, according to the Kempton article.

The exposé of alleged impropriety within the Pulp Workers stirred the executive board to legal action, and a partial retraction was secured from Kempton. However, the retraction did not touch upon the main points of the story, namely, the exploitation of the membership. Although there was never any evidence produced to confirm the charges made against Vice-President Tonelli, and though there was never any material to indicate Tonelli had benefited from his alleged actions in respect to Local 679, "The Dowry" was to become the basis for a campaign against him.

Two members of the Pulp Workers staff had seized upon "The Dowry" as a means of discrediting Tonelli. International Representatives Paul Hayes and Pat Connolly had collaborated in 1959 at the convention in a challenge to the hierarchy, and they utilized the Kempton article to publicize the alleged activities of Vice-President Tonelli. They mailed copies of "The Dowry" to Pulp, Sulphite locals throughout the United States and Canada, and demanded an investigating committee look into the affairs of Tonelli. At the meeting of the executive board in January 1960, action was taken:

1. *New York Post*, December 4, 1959, p. 46.

> It was thought adviseable to appoint a committee of three members to make a quiet investigation of these topics, rumors, and inuendos that have been circulated against Vice-President Joseph Tonelli at the Montreal Convention and continued to be circulated after the Convention.
>
> We appointed three outstanding members of the Executive Board to serve on this Committee. Vice-President Ralph W. Leavitt was appointed Chairman. International Treasurer Frank C. Barnes was appointed Secretary, and Vice-President Lewis [sic] H. Lorraine was appointed as a third member of the Committee.[2]

The investigating committee was to probe the stories and rumors that circulated about Joe Tonelli at the Montreal Convention and was further charged with investigating the allegations of Murray Kempton in "The Dowry."

> The rumors and charges made against Vice-President Tonelli at the Convention in Montreal came about as a result of a political campaign where there was a contest for the office of First Vice-President of the International Union; however, Brother Tonelli was the Third Vice-President. Some of these rumors and charges were as follows:
>
> 1. That Joseph Tonelli was connected with the New York underworld.
> 2. That shortly before the Convention, two New York gangsters visited the office of one of the delegates, locking the door behind them and threatened the life of that delegate unless that local union supported Joe Tonelli for President of the International Union at the Convention.
> 3. That some of the delegates at the Convention were being followed and even threatened.
> 4. That a planeload of New York "goons" were coming to Montreal.[3]

The Committee gave its report at the executive board meeting in April and found that the rumors had no basis in fact.

In regard to the charges of Murray Kempton, the investigating committee stated:

> Vice-President Tonelli was questioned on all the subjects mentioned above in this report, as well as many others which have not been mentioned specifically. In all the questioning, he talked freely and frankly on all subjects. His statements and answers to direct questions were given freely and unhesitatingly.[4]

The committee concluded:

> It is the unanimous opinion of this Committee that the various stories and rumors originating in the City of New York and designed

---

2. John Burke to George Lambertson and Maurice Albert, August 22, 1960. Papers of the International Brotherhood of Pulp, Sulphite and Paper Mill Workers, Rank and File Movement for Democratic Action, Wisconsin State Historical Society, Madison, Wis.

3. International Brotherhood of Pulp, Sulphite and Paper Mill Workers, "Minutes of the Meeting of the Executive Board, April 1–3, 1960," p. 6.

4. *Ibid.*, p. 7.

> to discredit Vice-President Tonelli are due mainly to personal differences between certain individuals and the vice-presidents in the area. Many of the differences and animosities are the result of organizing activities in the area resulting in jurisdictional problems. Due to the Vice-President's responsibilities in organizing work and the unique situation in New York City, the Committee cannot accept the conclusion that Vice-President Tonelli can rightfully be charged with the major part of the responsibility for this condition.
>
> This Committee was instructed to make a preliminary investigation and report back to the Executive Board. The Committee found nothing in this investigation that would warrant further investigation.
>
> It is the unanimous opinion of the Committee that Vice-President Tonelli has done an outstanding job for the workers in New York City, and for this International Union, under the most trying and difficult conditions. No doubt he has made some errors in judgment during the more than fifteen years he has served as an International Vice-President. Is there any other member of the Executive Board who can say he has not made any mistakes in judgment during his term of office? A man should be judged by the overall contribution he has made to the labor movement and the workers in the industry.
>
> Those persons who have been parties to the circulation of the rumors against Vice-President Tonelli, in the judgment of the Committee, have contributed more to undermining the good name of the International Union than Vice-President Tonelli has by any of his activities.
>
> The Committee is fully aware of the constitutional right of any delegate to seek and be a candidate for office at any International Convention. The Committee, further, is aware that in the heat of a political campaign statements are often made which are not entirely ethical. However, rumors or unsubstantiated charges intended to blacken the name of an International officer or other candidate and cast reflection on the entire executive staff or the International Union, are inexcusable. Any candidate for office who uses such tactics or, who knowingly permits those who sponsor or support his candidacy to use them in his behalf, should be called to account for such action by the International Executive Board.[5]

The committee may have been setting the stage for the dismissal of Representatives Connolly and Hayes. At the July 1960 meeting of the executive board, the two representatives were summoned and given a chance to present their charges against Vice-President Tonelli. After the sessions with Connolly and Hayes on July 6, 7, and 8 in Glens Falls, New York, President Burke appointed a three-man committee to prepare a report of the activities of the two men. Vice-Presidents Oliver, Parker, and Hansen were the members. The Special Board Committee reported that Connolly and Hayes:

> appeared before the Board and read prepared statements, with oral explanations, regarding rumors of misconduct by Vice-President Tonelli. After carefully reviewing their statements and the report of

5. *Ibid.*, p. 8.

> the Special Committee, with the complete absence of any documentary evidence, and because Vice-President Tonelli openly and satisfactorily answered all questions asked of him by the Board, and effectively refuted all the allegations made against him, the Executive Board finds that all of the alleged rumors made to date are unfounded and unwarranted. Therefore, in the total absence of any proof, the Board holds Vice-President Tonelli blameless of the allegations made against him.[6]

At the same session of the executive board, Connolly and Hayes were fired. The vote on the discharge motion was eight in favor, five opposed.

While the report of the Special Investigating Committee was signed by all three members, one of them, L. H. Lorrain, had second thoughts on the matter. On May 10, 1960, Lorrain wrote to the committee secretary, Frank Barnes, commenting:

> I am not in complete agreement with you that the report, as you say "represented the facts in the case." This investigation was in reality somewhat limited. Some very important witnesses . . . refused to appear before the Committee, and there was no interview of any kind with Murray Kempton.
>
> I have reread the report, recently, and I am very much disturbed to see how much weight we gave to the fact that corruption was rather common in New York City, as though this excused corruption in us. I would be very hesitant about having the report widely distributed throughout the Union, as I think it makes our Executive Board look somewhat less concerned about these problems than our membership expect us to feel. For these reasons, I have withdrawn my name from the report.[7]

Lorrain eventually went along with the majority and signed the report, but not without the misgivings expressed in this letter to Barnes.

## *The Program for Militant Democratic Unionism*

At the same time the executive board was terminating the services of Representatives Connolly and Hayes, reformers were readying their proposal for changes within the International Union. Their suggestions followed the line of thought that had been developed at the conventions in the 1950s, but with considerably more detail. A complete plan of reform was drawn up, entitled *A Program for Militant Democratic Unionism*. Local 312, Ocean Falls, British Columbia, apparently drafted the Program and on July 12, 1960, R. B. McCormick of that local sent President Burke a copy with a covering letter. McCormick noted that over the years some members of the International had attempted to secure changes in the Constitution which they felt would be beneficial to the union.[8]

---

6. "Minutes of the Meeting of the Executive Board, July 6–8, 1960," pp. 7–8.
7. Papers, L. H. Lorrain to Frank C. Barnes, May 10, 1960.
8. *Ibid.*, R. B. McCormick to John Burke, July 12, 1960.

The Program was lengthy and detailed a long list of proposed reforms. These included:

1. Secret ballot election of all officers.
2. Regional election of vice-presidents.
3. International Convention every two years.
4. Four officers to be elected by convention delegates: President, executive vice-president, secretary-treasurer, director of organization.
5. Separation of office of President-Secretary when John Burke left office.
6. Mandatory retirement of officers and staff at age 65.
7. Abolition of the office of auditor and selection of a Certified Public Accountant to audit the books of the International.
8. More militant collective bargaining.
   a. An end to private understandings and back door deals.
   b. Absolute prohibition against any representative accepting a gift or favor of any kind from employers, including drinks or meals. Conflict of interest to be eliminated.
9. Greater recognition of the role of skilled workers.
10. Formation of a Canadian Council within the International to develop plans and policies for collective bargaining in Canada.
11. Ultimate merger with the International Woodworkers of America.[9]

Introduction and circularization of the Program coincidentally with the discharge of Representatives Connolly and Hayes created much excitement within the ranks of the International. President Burke replied to the reform supporters by citing the record of the past. He wrote one staunch reformer, Burt Wells of Local 242, Rose City, Portland, Oregon:

> You must know our union is making progress in all activities. We are organizing the unorganized, we are signing agreements with more and more companies. Wage rates in the plants are the highest they have ever been, fringe benefits are the best they have ever been. Our members have more security than they have ever had. In view of all this, we have a great union. Our union has a record of accomplishments and the members have reason to be proud of it.[10]

Burke undertook to reply to the Program point-by-point in a letter to McCormick on July 16, 1960. He noted that many of the changes suggested in the Program had been suggested at conventions over the years and had been voted down by the delegates. He objected to the Program on the grounds that it would raise the costs of operating the union. He concluded:

9. *Ibid.*, Only the most significant sections of the program are reproduced here.
10. *Ibid.*, John Burke to Burt Wells, July 19, 1960.

> Do you think that if this International Union is full of corruption we would have money to give what we gave to the Woodworkers when they were on strike, what we gave to the Steelworkers that was returned, what we have given the Mine, Mill, and Smelter Workers and what we have given and are giving to various unions that have asked for help? I wish you could meet the officials of some of these unions, ask them what they think of the International Brotherhood of Pulp, Sulphite and Paper Mill Workers. Do you think you or anyone else could make them believe that this is a corrupt union.
>
> I resent very much anyone referring to this union as corrupt, as a corrupt organization. It is an idealistic union, it is a union interested in the cause of labor and humanity.[11]

The words of the International president-secretary did not seem to carry much weight with some of the rank and file members. As word spread of the discharge of Connolly and Hayes local unions began protesting to headquarters at Fort Edward. Requests for information were received from all areas of the United States and Canada. There soon developed an informal communication network among individuals who were interested in reform. Sentiment for change was concentrated on the West Coast of the United States and Canada, but there were supporters of reform on all parts of the continent. In July 1960, an Eastern Conference of Pulp and Paper Mill Unions was formed with Canadian and American co-chairmen. Pat Connolly wrote Angus Macphee: "It is my thought if a similar western council could be established, also a central council and a southern council, it would then develop the liaison to get ourselves in a pretty good position to do the job at the next convention."[12] Macphee replied:

> As I see it, and I believe it is also the opinion of others in British Columbia, the Program is the instrument, is the important issue, the firings are merely the instrument by which we can popularize our opinion.
>
> The present situation makes it an excellent time to get the Program out. Delays could be very costly.[13]

At the same meeting of the executive board where the action to fire Hayes and Connolly was taken, Treasurer Frank C. Barnes, a veteran of organizing efforts in the 1930s submitted his resignation because of ill health.[14] In view of the discussion that had occurred at the convention in 1959, Barnes's action seemed to confirm the suspicions of the reform group which felt it likely the opportunity would arise for the board to appoint a successor to the treasurer's post. Richard Ameden of Local 183,

11. *Ibid.*, John Burke to R. B. McCormick, July 16, 1960.
12. *Ibid.*, Pat Connolly to Angus Macphee, July 18, 1960.
13. *Ibid.*, Angus Macphee to Pat Connolly, July 27, 1960.
14. International Brotherhood of Pulp, Sulphite and Paper Mill Workers, *Pulp, Sulphite and Paper Mill Workers Journal*, July-August 1960, p. 1.

Everett, Washington, wrote Burke and stated it seemed to him the executive board was running a closed shop, eliminating competition for officers and frustrating the will of the convention delegates.[15] Burke replied that Barnes had to resign because of ill-health upon the advice of his physician. There was a contest for the office of treasurer, with Henry Segal winning by 8 votes to 5 for Charles Stewart.

Support for the reform program was building rapidly during the summer of 1960. The main advocates of change within the International were in frequent correspondence and were bolstering those of their acquaintances who wavered in the drive for reform. On July 30, a letter was written which seemed to express the sentiments of many Westerners. Herb Somes wrote Burt Wells: ". . . nothing would please me more than to see our own International degraded, insulted, and stomped on as we, the elected delegates of our own unions were when we attended the last wage conference."[16]

President Burke was literally bombarded with inquiries concerning Tonelli, Connolly and Hayes, and Frank Barnes. Local 168, for instance, wrote to him on August 10 inquiring about the disposition of the case against Vice-President Tonelli as made by the *New York Post.* They also wanted to know the reasons for the dismissal of Connolly and Hayes, and questioned the propriety of having the man who had been keeping the union books promoted to audit those same books.[17] The vice-presidents around the United States and Canada were getting letters similar to those being received at the headquarters in Fort Edward.

In August 1960, the Eastern Conference of Pulp and Paper Mill Unions addressed a questionnaire to President Burke that was more like a condemnation of the International Executive Board than a request for information. The questionnaire was prefaced:

> As the elected representatives of dues paying members of the membership we are insisting that President-Secretary Burke answer the following questions:
>
> 1. Do President Burke and other members of the Executive Board consider that in the positions they hold they are an entity separate and apart from the rank and file members?
> 2. Why didn't the Minutes of the January, 1960 Board meeting report that a Special Investigating Committee had been formed to investigate the Tonelli affair?
> 3. Who are the Board members who conspired to keep this fact from the rank and file?
> 4. Was there a motion made at the Board meeting that the International Union employ an impartial investigating agency to investigate the matter?

---

15. Papers, Richard Ameden to John Burke, July 20, 1960.
16. *Ibid.*, Herb Somes to Burt Wells, July 30, 1960.
17. *Ibid.*, Graham Mercer to John Burke, August 10, 1960.

5. Who made the motion?
6. What members of the Board favored the motion? What members opposed the motion and why?
7. Do you and the other members of the Board expect the rank and file to believe that Board members are capable of investigating Board members?
8. Did not David Ashe appear at the Board session as counsel for Tonelli? Is it not the same David Ashe who later appeared as counsel for the International Union in the matter of the Murray Kempton retraction?
9. At the Friday, April 1st, 1960 afternoon session of the Board . . . when you called upon Vice-President Leavitt, Chairman of the Investigating Committee, did he not have something to say in connection with the part he claimed that you, President Burke, had played in the activities of Joe Tonelli? Did he not in fact accuse you of directing Joe Tonelli to do in most instances what he had done?
10. Did he not accuse you of bringing Pat Connolly into the Fort Edwards office to do a job on Joe Tonelli?
11. What was your reply to these charges by Leavitt?
12. Did not Joe Tonelli threaten to blow the lid off the whole rotten mess if the Board members didn't get off his back? Did he not make accusations against you, President Burke? Did you not deny them by calling him a liar?

George Lambertson and Maurice Albert, Co-chairmen of the Eastern Conference, went on to question Burke about alleged inconsistencies in the minutes, and they also raised questions about Vice-President Tonelli's conduct. They wrote:

> Nor do the Minutes show that Tonelli admitted that he is a close personal friend of Alex Smalley, Vice-President of St. Regis Paper Co. and socialized with him at the race tracks and elsewhere. Nor do the Minutes show what explanation Tonelli gave for his ownership of the Tower Corporation, his relationship with the Cardinal Insurance Company fraud, and the Mastro Plastics Company Case; his signing of a forty-five month contract of Finishers, Inc., Cambridge, Mass. which provided no wage increase whatsoever over all that period of time with base rate of $1.00 per hour. What exactly did the Board ask Tonelli to explain, if anything? Did the Board ask him to produce his income tax return? Did the Board ask him to show how he came by $100,000 in 1950? If so, did he show proof of how he came of it? Is it not a fact that Joe Tonelli gave only vague and general explanations and furnished no documented proof that what he was saying was true?
>
> Vice-President Henry [sic] Lorrain reports that he was threatened by Francis Tierney at the July Board meeting to line up or else he would be knocked off the Board at the next convention. What about this? How many other Vice-Presidents were threatened in a like manner with better results than were obtained from Lorrain?
>
> The reports given by certain Vice-Presidents of what has happened at these three Board sessions told quite a different story than was told in the Minutes, and the failure on your part, President Burke, to give honest answers to these questions will serve only to convince the rank

> and file members of their suspicions that a majority of the Board members are attempting to cover up a major scandal within this International Union are well founded.[18]

The letter showed that there were many unanswered questions in the minds of the rank and file. These questions had to be answered before harmony could be restored to the ranks of the International, and President Burke made the attempt. In a reply to Albert and Lambertson which was almost as long as their letter to him, Burke attempted to deal with their questions. After a recital of his union activities since 1902, and a history of the strike at the International Paper Company of 1921, Burke began his reply. He noted the International Constitution required the executive board to manage the affairs of the union between conventions, and it was under that provision that the Special Investigation Committee was established. The committee report was not recorded in the minutes because:

> . . . we did not want this material in the hands of the enemies of our Union so that they could use it to work against our Union.[19]

According to Burke, no member of the board conspired to keep the report from the membership.

Burke's letter also noted that Murray Kempton had retracted his article, even though the *New York Post* had made a complete investigation of Vice-President Tonelli's affairs. In regard to the questions about the committee report on the activities of Vice-President Tonelli, Burke noted that Tonelli's friendship with Alex Smalley dated from the days when Smalley was an international representative for the Electrical Workers. Burke also presented the record of the union's dealings with Finishers Incorporated to refute the contention that Tonelli had signed a substandard contract.[20] The evidence indicated the Finishers' contract was not substandard. In regard to other questions about Tonelli, Burke stated the executive board did not ask Vice-President Tonelli to explain any of his affairs; he was not asked to produce his income tax return and he denied having $100,000 in 1950. Burke said he thought "that Vice-President Tonelli was very frank in giving explanations and answering questions."[21]

In reply to the question that alleged that Lorrain was coerced to sign the Investigating Committee Report, Burke quoted a letter from Lorrain.

> I wish to state that this report is untrue, I have never been threatened by anyone since joining this International. My associations with

18. *Ibid.*, George Lambertson and Maurice Albert to John Burke, August 1960. (Date illegible.)
19. *Ibid.*, John Burke to George Lambertson and Maurice Albert, August 22, 1960.
20. This author has the complete file of material on Finishers, Inc., including the original contract. To all evidence, the contract was standard. Wage increases were provided for, and while the base rate was $1.00, only two jobs had that rate. Also, $1.00 was apparently the common base rate for converting shops in the Boston area.
21. Papers, Burke to Lambertson and Albert, August 22, 1960.

> all the members of the Board, our Executive Board, have been pleasant and friendly. I have always been treated with utmost courtesy when I visited the International office.[22]

As usual, Burke concluded his letter by citing the proud record of the International down through the years.[23]

Response to the reform groups was not slow in coming from other quarters as well. The officers of Local 318 in New York City wrote McCormick on September 16, 1960, discussing the allegation of corruption in the New York area. They said, "There is not one bit of corruption in our Local Union," and according to the local officers the Program of Militant Democratic Unionism sought to "set members against locals, and locals against the Executive Board of this International Union."[24] They noted that the affairs of the unions in New York were closely supervised by various city, state, and federal agencies, and they offered to have McCormick come to New York City and inspect the books of the local. They concluded with a discussion of the Kempton column:

> Did you know that the Liberal Party, in which Vice-President Tonelli is an officer, lashed out at Kempton for his column. Did you know that even the Association of Catholic Trade Unions had to admit it could not find any corruption in our Union after its group conducted its own investigation?[25]

## *Developments in the Fall of 1960*

In the early fall of 1960, inquiries about the Program for Militant Democratic Unionism began coming in from all over the country, reflecting dissatisfaction with the International and seeking information about the reform movement. Burt Wells assured a correspondent in the South that nobody was attempting to ram the Program through, that it would have to stand the test of diversified interests. In the same letter he referred to photostats of important correspondence "which proves cases of sellouts by the International to various managements, proof that the economic report prepared by our Research and Education Director, George Brooks, is not getting into the hands of our Union negotiators through the action of the Executive Board."[26]

In a letter to the president of an eastern Local, Wells summed up the state of the reform campaign at that early date:

> This is a fundamental struggle for the rank and file. Those that think about their Union, those who are not informed of all the facts,

---

22. *Ibid.*
23. *Ibid.*
24. *Ibid.*, James Dassaro and Daniel Pancila to Robert McCormick, September 16, 1960.
25. *Ibid.*
26. *Ibid.*, Burt Wells to James Lallement, September 17, 1960.

> are torn between loyalty to their organization or to deviate from the principles of democratic unionism.[27]

By the middle of October 1960, Burt Wells was able to report a list of 140 interested reformers, and the establishment of a periodic publication, *Voice of the Program,* to spread the reform ideas throughout the continent.

In the early fall of 1960, the main thrust of the reform effort was directed at the charges of corruption in Local 679 in New York City. The West Coast locals were particularly concerned with the Local 679 situation. Local 433, Vancouver, British Columbia, the home base of Orville Braaten, wrote to Burke asking ". . . why has this man been allowed to operate the way he has for these past number of years."[28] By the time Local 433 was writing to Burke, however, the problem was on its way to resolution, at least from the standpoint of the International Union. Burke was able to write on November 8 that the Local and its president "have been problems to the International Union ever since they became a part of our organization. Anyway, Barbaccia has left us and so we will no longer be expected to do something about Local 679. Barbaccia has done the job for me."[29] According to a telegram from Barbaccia to Burke, the local membership had voted to disaffiliate from the Pulp Workers because they could not afford to pay the per capita tax.[30]

## *The George Brooks Affair*

While the reform effort was developing and building its strength throughout the fall of 1960, it was provided with another piece of ammunition to use against the International in January 1961. George Brooks, the director of research and education for the Pulp Workers, a nationally known figure in the field, resigned his position with the union.[31] A sheet headed "Resignation or Discharge" set forth an account of the termination of Brooks's employment with the International.

> In an informal session of the International Executive Board of which no minutes were kept the following events took place: One. Joseph Tonelli, Third Vice-President made a motion that the Board demand or require the resignation of George W. Brooks, Research and Education Director. This motion was seconded by Ivor D. Isaacson, Fourth Vice-President. Two. The outcome of the vote was as follows: Those voting for the retention of George W. Brooks were President John P. Burke, Vice-President Godfrey Ruddick, Vice-President Louis Lorrain,

27. *Ibid.,* Burt Wells to Richard Fontaine, September 28, 1960.
28. *Ibid.,* A. K. Stelp and M. Carter to John Burke, October 20, 1960.
29. *Ibid.,* John Burke to A. K. Stelp and M. Carter, November 8, 1960.
30. *Ibid.,* Telegram from Anthony Barbaccia to John Burke, October 6, 1960.
31. Brooks's resignation evidently occurred under pressure from the International Executive Board.

Vice-President Lloyd Oliver. The remaining eight Vice-Presidents and Treasurer Henry Segal voted in favor of the motion. Three. At this point George Brooks was summoned before the Executive Board and informed of the action taken. Four. Mr. Brooks accepted the mandate of the Board with the following provisions:

a. That he receive two years severance pay commencing April 1, 1961.
b. Pension protection.
c. Immunity for his staff in Washington.

The Executive Board agreed to the above terms with the understanding that the resignation would become effective April 1, 1961.

The following charges were offered in support of the motion made by Vice-President Tonelli during the informal session:

1. George Brooks was charged with exchanging correspondence without submitting a copy to President Burke in the following instances: A letter to Vice-President Lorrain criticising the Investigation Committee of the Connolly, Hayes, Tonelli affair. A reply to Frank Grasso's [then a General Vice-President of the UPP] circular entitled "A Declaration of Principles."
2. At the request of Patrick Connolly, George Brooks furnished him with the name of a firm that made lithographic reproductions of photographic prints. Connolly's purpose in this was to acquire reprints of a picture he had of Joseph Tonelli's mansion.
3. Brooks sent a letter to Connolly and Hayes expressing his opinion that the wage increase granted to the International staff in 1946 was illegal at that time. This was later made legal by the delegates assembled at the 1947 International Convention. In the same manner that the transfer of some five hundred plus dollars to the International Pension Fund was made legal subsequently by the action of the delegates at the 1956 Convention.
4. A statement by George Lambertson that George Brooks telephoned him asking that he get information on the Connolly-Hayes affair to the West Coast locals. Brooks flatly denied this and the only substantiation Lambertson could offer was a memo from his desk. [George Lambertson, President of Local 75, Berlin, New Hampshire, as Chairman of the Eastern Conference led the drive in that area for the reinstatement of Connolly and Hayes. For reasons unknown to the other members of the Eastern Conference Committee, Lambertson rejected his strong convictions in this matter and completely reversed his position.]
5. George Brooks was criticized by Vice-President Isaacson for taking upon himself the authority to inform the manufacturers of the delegates and the International Representatives' unanimous decision to file an unfair labor charge. This was in relation to the refusal of management to bargain pensions at the Association level during the 1959 wage conference.
6. Vice-President Isaacson, confirmed by Hansen, stated that George Brooks had remarked during the pre-wage conference in 1959 that if the mechanics demands were not met, they should seek other representation. These remarks were triggered by a survey of mechanics rates authorized by the pre-wage conference of 1958 which was compiled by the Department of Research and Education. This survey showed that the mechanics were fifty cents under the national level or average and at this conference they were seeking a

> twenty-five cent increase. This may have been a new concept to Brother Isaacson, but all delegates to every wage conference since the dawn of time in the ULA have toyed with the notion of seeking other representation.[32]

Reaction to the ouster of the research director was immediate and predictable. The reform elements were outraged. In the first weeks and months following the January board meeting, International Headquarters was flooded with letters demanding Brooks's reinstatement and vilifying the board members who had voted to ask Brooks to resign. Burt Wells mimeographed a "Bulletin" on the ouster of the research director.

> One of the blackest records in the history of the Union was chalked up on the weekend of January 14 and 15, 1961 when the International Executive Board pushed through the discharge of Education and Research Director George W. Brooks.
>
> In reviewing some of the actions of the Executive Board during the past year, there is only one conclusion that conscientious members of this great union can resolve. Our Union is being attacked from within by a power machine that is far too eager to take over by fair means or foul. The big push now is the elimination of any individual or movement that voices change or deviation from the decadent and antiquated status quo.[33]

Protests were received from all corners of the continent demanding an explanation of the resignation and seeking Brooks's reinstatement. The vice-presidents who had voted to oust Brooks were also put under considerable pressure by the locals they serviced. For example, Vice-President Ralph Leavitt received telegrams from Locals 41 and 61 protesting the action of the executive board and demanding the reinstatement of the research director.[34]

Local 375 in Philadelphia, Pennsylvania, was one of the Eastern locals that strongly supported the reform movement, and upon learning of Brooks's discharge, they telegraphed a protest to President Burke.[35] Burke replied to the local that the executive board requested the resignation of Brooks because of friction that had developed between the research director and some members of the executive board made it impossible for the board to work with him. The research director was given two years salary to permit him to do research and write a book.[36] Protests from the West Coast came from many locals. Burt Wells telegraphed an inquiry asking the vote for Brooks's discharge and the reasons for the action. The telegram concluded: "Drastic action by this local pends upon immediate

32. Papers, "Resignation or Discharge," mimeographed sheet.
33. *Ibid.*, "Bulletin," January 1961.
34. *Ibid.*, Telegram from Richard Currier, Local 61, and Leo Savage, Local 41, to Ralph Leavitt, February 3 and 10, 1961.
35. *Ibid.*, Telegram from Henry Rogers to John Burke, January 21, 1961.
36. *Ibid.*, John Burke to Henry Rogers, January 22, 1961.

answer."[37] Burke replied with his explanation about friction with the executive board.[38] The locals on the West Coast turned to the vice-presidents in the area for explanation of the executive board action. They were not satisfied. Melvin Melton telegraphed Burke that they had had a session with Vice-President Parker and he failed to advance a good reason for discharge.[39]

The main supporter of the reform effort on the executive board was Vice-President Godfrey Ruddick, who worked mainly in the southwestern part of the United States. It was Ruddick who supplied the details on the discharge of George Brooks, and in early February 1961, he circulated a letter to all Pulp, Sulphite Locals in the Southwest. He said he voted against the motion to request the resignation of Brooks because he felt the penalty was too severe for Brooks's alleged misconduct. Ruddick stated that the contributions of Brooks to the union outweighed whatever complaints the executive board had against him, and his services should have been retained.[40] The ouster of the research director divided the union into pro- and anti-Brooks forces. Even the International representatives became involved in the dispute. Shortly after Vice-President Ruddick's letter to the locals, the representatives working in his region wrote to President Burke requesting the reinstatement of George Brooks.[41]

It was apparent to the reformers, from President Burke's steady reiteration of his position on the resignation of George Brooks, and his unwavering position in opposition to some of their aims, such as regional election of vice-presidents, that some more organized method of presenting the reform position was needed. It was evident that telegraphic and mail protests were not changing the attitude of any of the members of the executive board. In February 1961, the first halting steps towards increased coordination of the reform efforts were taken. The infrastructure of the unions on the Pacific Coast provided a ready means for the advocates of change to popularize their views. At a meeting of the Pacific Coast Council of Pulp and Paper Mill Workers in February 1961, the facts about the discharge of George Brooks were widely disseminated and a letter of protest was sent to President Burke:

> The Pacific Coast Council has heard Vice-Presidents Isaacson, Parker and Hansen attempt to explain the charges that led to the resignation of George W. Brooks. The Pacific Coast Council strongly contends that these charges were entirely without merit and were not

37. *Ibid.*, Telegram from Burt Wells to John Burke, January 21, 1961.
38. *Ibid.*, John Burke to Burt Wells, January 23, 1961.
39. *Ibid.*, Telegram from Melvin Melton to John Burke, January 24, 1961.
40. *Ibid.*, Godfrey Ruddick to all Southwestern Pulp, Sulphite Locals, February 1, 1961.
41. *Ibid.*, Charles Stewart, W. A. Sims, Amos Lindsay, Shelby Phillipps, Wayne Glenn, Joseph Bradshaw, Durwood Ruddick, and Raymond Bradford to John Burke, February 7, 1961.

> worthy of the action that followed. This is to inform you that this Council wishes to be placed on record advocating the reinstatement of George W. Brooks as Director of Research and Education.[42]

A meeting of the Northwest Committee for Union Justice was called for February 20, 1961, at Portland, Oregon. Wells was counting on a number of people to attend. He listed Dick Ameden, Melvin Melton, Graham Mercer, Murray Randall, Bob Jordan, Glenn Munsey, Bill Perrin, and Harold King.[43] At the same time that the Western locals were striving to develop some coordinated campaign, the Eastern locals were thinking along the same lines. Leo Savage, president of United Gilman Local 41, Gilman, Vermont, wrote to Orville Braaten:

> We are quite concerned over one thing at the present and that is that we need someone to coordinate this movement and give it the proper leadership. We do not think that we can win this battle without a nation-wide conference and [sic] lay the groundwork to bring this Board into line.[44]

## *Formation of the Rank and File Movement for Democratic Action*

The desire for more coordination and leadership manifested in the February 20, 1961, meeting of the Northwest Committee for Union Justice and in the feelings of Leo Savage culminated the next month at a national conference in Colorado. At the Shirley Savoy Hotel in Denver on March 23–24, representatives from locals from all over the United States and Canada assembled to discuss the reform effort.[45] Leo Savage, United Gilman Local 41, Gilman, Vermont, was named temporary chairman, and Angus Macphee, Local 708, Prince Rupert, British Columbia, was named temporary secretary. Committees were appointed to deal with the various matters before the group, such as organization, name, structure, program, and finances. It was decided to establish an organization to be known as the Rank and File Movement for Democratic Action. It was

---

42. *Ibid.*, the Pacific Coast Council of the International Brotherhood of Pulp, Sulphite and Paper Mill Workers, affiliated with the Pacific Coast Pulp and Paper Mill Employees Association, Clarence W. Dukes, secretary-treasurer, to the executive board, February 1961. (Date illegible.)

43. *Ibid.*, Burt Wells to Walter Northrup, February 15, 1961.

44. *Ibid.*, Leo Savage to Orville Braaten, February 14, 1961.

45. *Ibid.*, "Minutes" of the March 23–24 meeting establishing the Rank and File Movement for Democratic Action. The representatives were: Local 375, Philadelphia, Penn., Henry Rogers; 68, Oregon City, Ore., William Perrin; 242, Portland, Ore., Burt Wells; 512, West Monroe, La., R. H. Chatham; 22, Fort Orange, N.Y., Harold Bartlett; 20, Glens Falls, N.Y., Mario Scarselletta, Jr.; 885, Frenchtown, Mont., Raymond L. Howell; 41, Gilman, Vt., Leo Savage; 713, Antioch, Calif., Clarence Dukes; 249, Antioch, Calif., Arthur Farrace; 183, Everett, Wash., Richard Ameden; 89, Kapuskasing, Ont., George Rosebush and James Ballantyne; 708, Prince Rupert, B.C., Angus Macphee; 433, Vancouver, B.C., Orville Braaten; 76, Powell River, B.C., C. M. Monat.

stated that "the primary aim of this movement is the reinstatement of Research Director George W. Brooks."[46] A preamble stating the purpose of the RFMDA was drawn up:

> The Rank and File Movement for Democratic Action of the International Brotherhood of Pulp, Sulphite, and Paper Mill Workers is opposed to those who would undermine the basic principles of free trade unionism. Its members, without fear, intend to take democratic action to defend these principles. They are opposed to any corrupt procedures or actions detrimental to the welfare of the membership of their International Union. They pledge their united efforts to safeguard the democratic character of the labor movement.[47]

RFMDA was to be financed by a monthly charge of five cents per member from each affiliated local. R. H. Chatham of West Monroe, Louisiana, was elected chairman, William Perrin of Oregon City, Oregon, vice-chairman, and Burt Wells of Portland, Oregon, secretary-treasurer. In keeping with the frequently expressed desire of some of the participants for regional election of officials, the RFMDA divided the continent into regions and elected an executive for each.[48] With this structure, the reformers were entering a new phase of their struggle. While the proponents of change were organizing to carry on their struggle, they did not neglect their missionary work. Local 89 of Kapuskasing, Ontario, circulated a bulletin to the Pulp Workers locals outlining its position on reform:

> Pat Connolly, Paul Hayes, George Brooks, who is next in the sights of some marksmen on our International Executive Board? ? ? ? Since our last International Convention these three men have either been discharged or forced to resign by the majority vote of our International Executive Board. In all cases President Burke, Vice-Presidents Ruddick, Oliver and Lorrain voted in favor of keeping these people. Does it not leave a sour taste in the mouths of our membership when men of the character of our President and the fore-mentioned three Vice-Presidents are overruled by the vote of the other members of the Board? We have asked repeatedly that these men be reinstated without loss of pay or seniority but once again this has been turned down by majority vote of the Executive Board. The full list of the staff of the International Representatives along with two members of the Executive Board in the Southern U.S.A. has signed a resolution for reinstatement of George Brooks. Locals in the New England States, the West Coast of the United States and the West Coast of Canada are of the same opinion. Local 89 lends its full support to reinstate these men. If we are going to have a democratic union with International Board Officers responsible to the membership things have got to be changed. Let President Burke and the other members of the

---

46. *Ibid.*
47. *Ibid.*
48. *Ibid.*, The regions and their officers were: New England, Leo Savage; Middle Atlantic, Mario Scarselletta, Jr.; Southeast, Vacant; Southwest, Vacant; Lake States, Vacant; Middle West, Vacant; California, Arthur Farrace; Northwest and Alaska, William Perrin; Ontario and Manitoba, George Rosebush; Western Canada, Orville Braaten; Quebec and Newfoundland, Vacant.

> International Board hear the voices of Newfoundland, the Eastern Canada Council and the Ontario and Manitoba Council of Pulp, Sulphite locals. Let us support President Burke and the three Vice-Presidents, Ruddick, Oliver, and Lorrain who had enough guts to stand up and be counted.[49]

The pressure on the vice-presidents for a satisfactory explanation of the forced resignation of George Brooks was particularly strong on the West Coast of the United States and Canada. On February 15, 1961, Vice-President Hansen had reversed his position at a meeting of Watson Island Local 708 in British Columbia. If Hansen's change of heart had persisted it would have represented a chink in the otherwise solid front of the executive board majority. But Hansen wavered, perhaps under the counter-pressure of his colleagues on the board. On March 15, "Stubby" Hansen, who had been elected to the board at the 1959 Convention by the action of the British Columbia locals, wrote Peter Marshall of Local 708:

> During my tenure as Representative and an elected Officer of the Executive Board I have without deterrent assumed and carried out my full responsibility in accordance with our constitutional qualifications. After full investigation, it is with profound regret that I withdraw my commitment made *in good faith* while attending the regular meeting of Watson Island Local 708 on February 15 last namely "in the event the majority of locals embracing the British Columbia membership at constituted convened meetings vote in favor in reinstating George Brooks I would support such an appeal," which I felt obliged to do at the time. This is one of the hardest and most distressing decisions I have been called upon to make during the past 25 years.[50]

That caustic critic of the International, Melvin Melton, who had helped fan the flames of discontent with his letter to the *Journal* in 1959, commented in 1961:

> During the past few years the International Executive Board has lost what little respect they had in this local union. Want to know some of the reasons? Here's a few. One, the highest per capita tax in the history of the International Union. Two, the lowest level of service to the local unions. Three, rampant stupidity, compounded semiannually. Four, lousy representation at wage conferences. Five, the illicit love affair with certain companies bordering on collusion. Six, the inability to accept, coupled with the unwillingness to face the present, to come out of the cocoon of the past. Seven, the paternal "poppa know best" attitude that abides no criticism. Eight, the eagerness to weed out anyone who shows the least trait of leadership.[51]

With the formation of the RFMDA the leaders of the reform group felt it was necessary to give their local membership some explanation of the ferment within their union. Graham Mercer of Local 169, Hoquiam, Washington, which had always been a thorn in the side of the Inter-

49. *Ibid.*, "Bulletin" dated February 27, 1961. Signed by James Ballantyne.
50. *Ibid.*, H. L. Hansen to Peter Marshall, March 21, 1961.
51. *Ibid.*, Melvin Melton to John Burke, February 17, 1961.

national, wrote to the membership of the local setting out the rationale for RFMDA:

> Our primary target is, of course, to give more power to the local unions and to take some away from the International. We are not yet concerned here with outside matters but with powers *within* the structure of the union. We feel that we *must* have more control over the men who represent us in wage negotiations and in other matters pertaining to our union. In order to do this we must have a different means of electing these men to office. If you'll read your International Constitution, you'll find that Vice-Presidents are elected by vote of the entire delegation to each Convention. This, to us is like having the entire country vote to elect the Senators from the State of Washington. We want to have this changed so that the Vice-President from this area will be elected by the people from this area and will then be beholding [sic] to them for his position. We feel that this one factor alone will make our officers more responsive to our desires. There are other changes which should be made in the bylaws, but this is probably the most important since the others will follow more easily if we can make this one.[52]

It seems from this letter that only a month after the Denver meeting establishing RFMDA, the aims of the reform group were changing. RFMDA's stated purpose was the reinstatement of George Brooks as research and education director. In this letter Mercer indicated that the aims of the reformers were considerably more than mere reinstatement. RFMDA was becoming an instrument to promote fundamental change within the International Union. The changes advocated by RFMDA were changes long opposed by the officials of the International, and they evidently felt compelled to take some action beyond writing letters to the reform group extolling the virtues of the Pulp, Sulphite Workers.

In April 1961, the International hierarchy attempted to take the offensive in the struggle with the advocates of change. The stratagem decided upon was a fact finding trip by some supporters of RFMDA to New York City so they could see the operation of the New York locals firsthand and report to the membership on the West Coast that the New York operation had a clean bill of health. RFMDA was well posted on the tactics to be employed by the International:

> Henry [Rogers], I received a call at my house Sunday and part of the conversation was about President Burke sending $798 to two members of RFMDA for fare to New York City. They are also to receive $20 a day for five days as pay plus tabs to be picked up by "the great white warrior" and leader, Joseph Tonelli. I have the two Brothers' names.[53]

The people brought to New York were Clarence W. Dukes and Arthur Farrace, both of Antioch, California. Dukes was from Local 713 while

52. *Ibid.*, Graham Mercer to membership of Local 169, Hoquiam, Wash., April 22, 1961.
53. *Ibid.*, Harold Bartlett to Henry Rogers, April 4, 1961.

Farrace came from Local 249. Both Dukes and Farrace made lengthy reports on their trip, though Farrace was sick for most of the time in New York. According to the report by Dukes, he and Farrace were sent checks to cover their first-class plane fare plus wages and twenty dollars per day to cover expenses. They were met at the airport by a representative from Vice-President Tonelli and were treated very well, eating at the most expensive restaurants. They were chaperoned at all times. They spent a considerable amount of time with Tonelli, who tacitly agreed that he was the strong man of the union.

> From Tonelli's remarks they inferred that he had frightened other staff members into voting his way at the January [1961] Board meeting and that they are still frightened for their jobs. Dukes stated that he was still with the movement for reform but that he would be willing to make a deal with Tonelli if we ultimately failed in our efforts to garner enough strength to accomplish the reforms. He made the proposition to Tonelli to get rid of Isaacson and this appears to be of paramount importance in his area. Tonelli was noncommittal as to specific details, but indicated that under certain circumstances he might go along with this line of thought, asking if they had someone to propose and intimating that he could make their desires possible if he wished.[54]

In spite of his illness, Art Farrace made a more complete report on the trip than his travelling companion:

> There is no love between Tonelli and Ruddick. I can't quite make up my mind but I think there is a power play between these two and I think that George Brooks was the fall guy. There is no question that Tonelli has the power to swing the Convention his way. I accused Tonelli of his powers and at first [he] denied that he was this powerful but after we explained that we know all of his position, he agreed that he could swing the Convention. He is opposed to rehiring Brooks because he felt that George interfered with union business when he should have been doing his job. This is Tonelli's excuse, I believe that Tonelli knows that Brooks carries a lot of weight with the rank and file and is a problem to him so he had to get rid of him. I wish I was smart enough to understand the politics that are going on in our International. I have the feeling that it's bigger than we could believe.
>
> I have a feeling between Tierney [the office manager] at the International Office and Tonelli, everything is pretty well in their control. We told Tonelli that we wanted Ike (Isaacson) out as President. We couldn't speak for you people in the North but we did say that you were very disappointed with your V.P. He asked us who we wanted to be V.P. in California and we suggested Ray Bradford and he wanted to know what we had against Cash Price. So you can see, this is in the wind. Personally, I am disappointed with Cash. Anyone who could go along with Ike couldn't possibly do a job for the membership. We got this commitment from Tonelli, that we in California could pick our V.P. if he agreed he would support him for election. So you can see that he knows that he has the power. . . . We told

54. *Ibid.*, Clarence Dukes's report on the trip to New York, April 1961.

> him that we wanted Ike to resign at the Convention and not before and that we wanted to elect our own V.P. by areas. He opposed this. He felt it was wrong to elect V.P.'s by areas. Well, we opposed his position but you can understand why he feels this way. He knows he has control. With area elections he would be finished as the power and there is no question in my mind that he has this power. I feel that we can do a job and get strong and well organized when we go to the Convention. We must get the "Program" in or all is lost and we must get the help of not only the rank and file but the inter-representatives. This is very important for they can swing a lot of votes and also can spend money which is very important at the Convention.[55]

Another step was taken in the spring of 1961 designed to counter some of the material the reform group was circulating. The RFMDA was distributing a newspaper, the *Amplifier.* To counter the statements in the *Amplifier,* there appeared in May 1961 a newspaper called *Truth.* The origin of *Truth* is obscure and it carried an unreadable printing trades label. The reform elements were warned in advance of the appearance of the paper. Henry Rogers advised Burt Wells of the imminence of the publication and asserted the paper was being prepared in Fort Edward. Rogers noted *Truth* would contain a "Hall of Shame" whose first occupants would be Connolly, Hayes, and Brooks.[56] As Henry Rogers had foretold, *Truth* did indeed appear in May 1961 and was clearly a paper on the side of the hierarchy in the developing contest. The first issue's lead story was headlined "Brooks Finished" and was written by someone operating under a pseudonym, Rudi Brooks. *Truth* also contained correspondence that the reform group had assumed was restricted to the hands of the executive board. The supporters of RFMDA were stirred to protest by the appearance of *Truth,* and Albert Sterman of Local 375 wrote President Burke:

> We are in receipt of an anonymous twelve page document titled *Truth.* It contains many letters and memorandums which we presume were only in the hands of [the] International Executive Board. Can you tell us how does it happen anonymous persons have access to these confidential communications? How is it possible for anonymous persons to obtain such communications from our International Union Office? How does our International Union condone anonymous publication? Who is financing this publication? We would appreciate your answers to the above questions.[57]

Burke replied to Sterman that he knew nothing about *Truth* and that it did not contain any confidential information.

> No. As I go through this publication I can see nothing in it that was very difficult for those who prepared it to obtain, and not from our files here at the International Office.[58]

---

55. *Ibid.,* Arthur Farrace to William Perrin, May 20, 1961.
56. *Ibid.,* Henry Rogers to Burt Wells, May 9, 1961.
57. *Ibid.,* Albert Sterman to John Burke, May 24, 1961.
58. *Ibid.,* John Burke to Albert Sterman, May 31, 1961.

Another participant in the controversy over *Truth* was Paul Hayes, who was still seeking reinstatement to his position as International representative. He wrote to Burke alleging that the editor of *Truth* was Francis Tierney, Burke's office manager. Hayes stated that preparatory work for the paper was done in the office of Treasurer Henry Segal, mainly when Burke was absent. Hayes alleged that Tierney, Tonelli, Leavitt, and Segal had held a strategy meeting and decided to distribute the paper.[59] Burke replied to Hayes at some length and with a certain degree of annoyance usually not found in his correspondence. Burke stated that while he had nothing to do with *Truth,* in general he agreed with its charges concerning the activities of Connolly and Hayes when they were charged with stirring up dissension within the International. He asked Hayes if it was not true that George Brooks wrote Pat Connolly on June 30, 1960, regarding salary increases Burke had put into effect in 1946 for the International vice-presidents and representatives. He also asked Hayes if it was not true that Brooks cooperated with Pat Connolly in securing a printer to reproduce a thousand copies of a picture of Vice-President Tonelli's house. Burke continued that when Connolly and Hayes were fired by the executive board the information was relayed to Brooks who called George Lambertson and "urged him to get busy in organizing the East Coast against the Executive Board. He assured George Lambertson that he controlled the West Coast." Brooks denied Lambertson's story. "However, after George Lambertson came on from Berlin, New Hampshire, and faced him, he changed his story and admitted having called him on the telephone from his cottage in Virginia." Burke continued to reiterate that he had no knowledge of the origin of *Truth.*[60]

The dispute over the persons responsible for *Truth* lasted through the summer of 1961, with members of the reform group writing Burke and asking questions of him. In a letter to Albert Sterman, Burke tried to end the discussion:

> Now I will try to answer your questions in regard to the publication *Truth.* You asked me if I had any knowledge whatsoever of who wrote this publication. No, I do not. Yes, I have talked with Francis Tierney, Joseph Tonelli, and Henry Segal. I have pointed out to them that they were being charged with having prepared and issued this publication. They denied it most emphatically. You ask if I made an investigation of this matter to satisfy myself that no funds of this International Union were used directly or indirectly to publish or distribute this publication. Yes I have. None of the funds of this International Union could be used for this purpose because I sign all of the checks issued by this International Union.[61]

59. *Ibid.,* Paul Hayes to John Burke, May 12, 1961.
60. *Ibid.,* John Burke to Paul Hayes, May 16, 1961.
61. *Ibid.,* John Burke to Albert Sterman, July 20, 1961.

The RFMDA was facing more fundamental problems than those presented to it by the actions of the executive board. Affiliations with the movement were lagging. Revolts must be financed and this one was in some difficulty. At the end of May 1961, Henry Rogers wrote to Godfrey Ruddick complaining about the lack of response in the South.

> It is with deep concern that I feel obligated to bring to your attention the matter of affiliation to the Rank and File Movement for Democratic Action. As you know, I keep in close contact with Burt Wells. In discussing the subject matter with him, I was shocked to hear that there are only seventeen local unions that are presently affiliated and paying per capita tax. It is wonderful to talk about a reform movement. We look toward the necessary reforms that must come, but how in the world are we going to do the job without money. Your Southern group can help us a lot. Only if somebody begins to push those local unions a little. I constantly brag about the solid South being a vital part of our movement, but if somebody checks on this statement they are going to be very disappointed because there is only one Southern local that is paying per capita tax to our organization . . . .[62]

The situation with respect to affiliations did not improve much in the following months. Rogers wrote to Wells:

> We are very sorry to hear you express your concern over the lack of affiliations. Of course, I don't blame you a bit. I am inclined to feel the same way and there are others who share your views on this matter.
>
> It is well over five months since our movement began and how far are we? It is a damn good thing that we are lucky to have you and a few others like you. Otherwise, this thing would stagger me.[63]

The RFMDA had set forth a program with only one feature, the reinstatement of George Brooks. As the months passed, it became apparent that more than the issue of reinstatement was needed to sustain interest in the program. The aims of the RFMDA were broadened throughout 1961 to include some of the changes its supporters felt were necessary in the conduct of the International. By the fall of 1961 Henry Rogers wrote:

> I agree completely with you that we cannot restrict the issue to George W. Brooks. As a matter of fact, it has been a long time since I have heard anyone suggest that this is necessary or even desirable. It gave us a starting point, that is all. I am sure George himself feels strongly about this. The program is being broadened at every area conference to include secret ballot election of International Officers, regional elections and many other reforms that you and I feel are necessary.[64]

The supporters of RFMDA also attempted to compare the activities of Brooks's successor as research director, John J. McNiff, with those of

---

62. *Ibid.*, Henry Rogers to Godfrey Ruddick, May 31, 1961.
63. *Ibid.*, Henry Rogers to Burt Wells, July 31, 1961.
64. *Ibid.*, Henry Rogers to Angus Macphee, September 8, 1961.

Brooks. They found particular fault with the service rendered by the department under McNiff and the expenses incurred by the new director. Early in 1962, Burt Wells initiated a campaign to question the services of McNiff, and he wrote to supporters of RFMDA asking that they write to President Burke and discuss the matter with him. The outline letter sent by Wells questioned the increase in the expenses of the Research and Education Department, while it was alleged that service had not been stepped up. Henry Rogers, too, questioned McNiff's expenses.

> As for John McNiff, there is no justification for his excessive expenditures traveling all over the country presumably to build and mend fences for the opposition. If McNiff is a new man and bound to have a lot to do, then why doesn't he stay in Washington, D.C., and learn his job and not travel 75,000 miles in eight months at our expense.[65]

President Burke constantly defended his new research director and portrayed his work in a favorable light in the Pulp Workers *Journal*. In the heat of the controversy, pro-RFMDA locals were completely opposed to McNiff. Writing to the editor of the *Journal*, Local 704 protested the credit given McNiff in pension negotiations with Kimberly-Clark: "Our delegates report that Mr. McNiff contributed nothing to the outcome of the meetings even though he was present."[66]

RFMDA also wanted a careful watch to be kept on expenditure of union funds. The Pulp Workers made regular financial reports to the membership, but some RFMDA supporters felt the reports concealed more than they showed about the disbursement of union funds. In September 1961, Henry Rogers wrote to Orville Braaten of his feeling that the item "Organization Expenses" in the reports concealed expenditures to defeat the aims of RFMDA. He enclosed a draft letter the auditor of the Pulp Workers wrote inquiring about a breakdown of the item.[67] Perhaps to RFMDA's surprise, Burke made an effort to go along with the request for more detail in the financial reports. He sent a breakdown of the "Organization Expense" figure to R. H. Ameden, covering the cost of organizing by state and campaign.[68]

The fight of RFMDA to change the union was not without defectors. George Lambertson, who had been a supporter of reform in New England, was the person who supplied the executive board with the material to make a case against George Brooks. Lambertson was later put on the

---

65. *Ibid.*, Henry Rogers to Burt Wells, February 21, 1962.
66. *Ibid.*, Lewis Donovan to John Burke, April 16, 1962. John McNiff, who replaced Brooks as Research and Education Director was the same person who had testified before the McClellan Committee that Local 679 exploited its Puerto Rican membership.
67. *Ibid.*, Henry Rogers to Orville Braaten, September 14, 1961.
68. *Ibid.*, John Burke to R. H. Ameden, October 24, 1961.

payroll of the International as a representative. The RFMDA commented through its newspaper, the *Amplifier,* about the appointment:

> This man appears to be a tool of Tonelli and Company. Serious questions have been raised in the Union about his trustworthiness. Is this the time to add him to the staff, in the interests of the welfare of the organization? We think not.[69]

More serious than the defection of Lambertson was the change of heart of one of RFMDA's founders and first chairman of its Middle Atlantic Region, Mario Scarselletta, Jr., of Local 20. On January 18, 1962, Scarselletta wrote Burt Wells and told him he was resigning as chairman of the Middle Atlantic States Region of RFMDA and that Hudson River Local 20 was disaffiliating from the movement. Scarselletta said he had talked to Vice-President Tonelli, who had answered all the questions put to him. Scarselletta was upset with the political position of some of the members of British Columbia. He said,

> Those to whom I make reference, and it is quite obvious who they are, in my opinion follow the Communist Party line without the slightest deviation. We of Local 20 cannot condone this and we refuse to be associated with individuals who would use our International Union for ulterior purposes.[70]

The attack on the political persuasion of some of RFMDA's supporters opened a new front in the struggle, a front where RFMDA was somewhat vulnerable. In the summer of 1961, two prominent supporters of the reform movement, Robert McCormick and Angus Macphee, both of British Columbia, had made a a trip to Cuba to celebrate the July 26 Revolution with the Castro government. Other RFMDA supporters from British Columbia had also been identified with espousal of allegedly left-wing causes. Orville Braaten was often mentioned in this connection. RFMDA had to answer the charges of Scarselletta. George Rosebush of Local 89, Kapuskasing, Ontario, wrote to Scarselletta:

> In your letter to Burt Wells you accuse some of the Brothers in British Columbia of being Communists. You did not mention any names or offer any proof of your convictions and I think this is highly unfair and absolutely untrue. It seems that if you cannot find anything else to call somebody and that you have no proof then all you have to do is brand that person or persons a Communist.[71]

In March 1962, the executives of RFMDA wrote Scarselletta to notify him he had been replaced as Middle Atlantic States chairman long

---

69. Rank and File Movement for Democratic Action, *Amplifier,* September-October 1961, p. 1.
70. Papers, Mario Scarselletta, Jr., to Burt Wells, January 18, 1962.
71. *Ibid.,* George Rosebush to Mario Scarselletta, Jr., February 15, 1962.

before he resigned and that his local had not been affiliated with RFMDA since May 1961, since the per capita tax had not been paid.[72]

The issue of communist influence within RFMDA was to recur from time to time, and the leaders of RFMDA were forced to spend time and effort to combat it. In a meeting at Mobile, Alabama, April 26–28, 1962, Vice-President Lloyd Oliver allegedly branded supporters of RFMDA as communists.[73] R. H. Chatham replied to Oliver:

> If Lloyd Oliver thinks that anyone in our movement is a Communist and is doing harm to this International Union why doesn't he name names and bring charges in writing? All we are is a group of local members who disagree with decisions of the Executive Board and have joined together to discuss and work for changes in the Constitution. Is not this our legal and just right? We resent our per capita tax being used to fight us.[74]

The communist charge remained and RFMDA continued to resort to the *Amplifier* to answer it. Graham Mercer of Local 169 wrote about the "Red Herring" of the communist issue:

> The attack is a simple example of the red herring. The goals of RFMDA are simple things:
>
> Regional election of vice-presidents.
> Compulsory retirement of officers at age 65.
> Greater financial disclosures and responsibility.
> Reinstatement of employees who were in the opinion of RFMDA improperly separated from the Union.[75]

While RFMDA attempted to defend itself against the "red herring" of communism, it was also attempting to broaden its appeal to increase its base of support for the convention in Detroit in 1962. Though the supports of reform still felt aggrieved at the discharge of George Brooks, Pat Connolly, and Paul Hayes, thoughtful members realized it was not enough to simply propose reinstatement at the convention. The program had to be expanded to include basic changes in the constitution and the manner in which the union conducted its business. To meet these challenges, Bill Perrin wrote to R. H. Chatham in early 1962: "I think the Executive should be called into session early in March and I think Denver at the same hotel will be satisfactory."[76] A meeting was indeed held in Denver, and out of the meeting came an expanded RFMDA with an ambitious ten-point platform of reform:

72. *Ibid.*, R. H. Chatham, William Perrin, and Burt D. Wells to Mario Scarselletta, Jr., March 23, 1962.
73. *Amplifier*, June 1962, p. 3. Letter from R. H. Chatham to John Burke, May 17, 1962.
74. *Ibid.*
75. Graham Mercer, "That Red Herring Again," *Amplifier*, July 1962, p. 1.
76. Papers, William Perrin to R. H. Chatham, January 11, 1962.

1. Regional election of Vice-Presidents with recall provisions.
2. Adoption of AFL-CIO Ethical Practices Code.
3. Compulsory retirement at age 65 (with exception of President Burke).
4. Establish office of Executive Vice-President.
5. Establish a Public Appeal Board.
6. Provision for calling special conventions.
7. Establish a more complete financial report.
8. Reinstatement of Paul Hayes.
9. Reinstatement of Pat Connolly.
10. Reinstatement of George W. Brooks.[77]

The revised program dropped the reinstatement of George Brooks from top to bottom priority, and from March until the convention in September, the RFMDA hammered at the theme of democratizing the union through regional election of vice-presidents. The program that emerged from the March meeting had much in common with the Program for Militant Democratic Unionism which had been introduced almost two years earlier.

The *Amplifier*, and other papers that reform supporters had established, such as the *East Coast Newsletter* published by Local 375 in Philadelphia, and the *Southeast Newsletter* put out by Locals 400 and 738, Brunswick, Georgia, and Wilmington, North Carolina, took up the cause of reform and gave it wide publicity. From March to September, the emphasis was on the issue of regional election of vice-presidents.

The *Amplifier* commented on the March meeting: "We are no longer in a revolt—we are engaged in a pitched battle provoked by the International oppositionists."[78] Typical headlines in the reform newspapers included: "This is a Plug for Regional Election of Vice-Presidents"; "Why we are for Regional Elections of Vice-President"; and "Regions Should be Defined in the Constitution." The *Newsletter* of the Northwest Committee for Union Justice, the regional affiliate of RFMDA in the Pacific Northwest, commented that the main reason they were for regional election of vice-presidents was to ensure democracy. The Northwest Committee felt that regional election, coupled with compulsory retirement, also an RFMDA objective, would protect the membership from the practice of having the executive board select vice-presidents to fill vacancies caused by death or resignation. The Northwest Committee presented the

77. *Ibid.*, "Minutes" of March 12–14, 1962, meeting at Denver, Colo. The item concerned with reinstatement of Pat Connolly was dropped by the date of the convention since Connolly had found a job in private employment and did not want reinstatement.

78. *Amplifier*, April 1962, p. 2.

record on selection of vice-presidents: Of the thirteen members of the executive board in 1962, six had been appointed by the board itself.

RFMDA stood on the record with a very ambitious program of reform. It expected to come to the convention with its ten-point program, and this caused some concern among its supporters. Henry Rogers from Philadelphia felt the program was too broad, and in June 1962, only two months before the September convention, he wrote:

> We stand on the record with a very ambitious program. But frankly I do not see how we can possibly push all of the items we talked about. I think we shall be doing very well if we can do a job on regional elections, compulsory retirement, the reinstatements, financial responsibility and one or two others. I don't even know where we stand on regional elections in terms of what resolutions we are going to support.[79]

By the time of the convention in Detroit in September 1962, most of Henry Rogers fears had been allayed, as RFMDA came to the convention with a well-reasoned plan for reform within the union.

79. Papers, Henry Rogers to William Perrin, June 12, 1962.

# 6

# THE DETROIT CONVENTION: Defeat of the Rank and File Movement for Democratic Action

## *Pre-Convention Activities*

The main efforts of the Rank and File Movement for Democratic Action were directed toward changing the constitution of their International Union. The Pulp Workers twenty-sixth convention was scheduled for September 10–16, 1962, in Detroit, Michigan, and was to provide the setting for RFMDA's effort to secure the reforms its members felt were necessary to democratize their union.

In preparing for the convention, RFMDA built upon the program that had been outlined in the Denver meeting in March 1962. International Representative Pat Connolly had secured satisfactory employment and showed no desire for reinstatement, so the final draft of the RFMDA program omitted this point. Otherwise, the program presented at Detroit was the same as that drawn up at Denver. The reform proposal embodied nine points:

1. Regional election of Vice-Presidents with recall provisions.
2. Adoption of AFL-CIO Ethical Practices Code.
3. Compulsory retirement at age 65 (with exception of President Burke).
4. Establish office of Executive Vice-President.
5. Establish a Public Appeal Board.
6. Provision for calling special conventions.
7. Establish a more complete financial report.
8. Reinstatement of Paul Hayes.
9. Reinstatement of George Brooks.

The Rank and File Movement had a keen appreciation of the task that lay ahead at the convention. They prepared for the undertaking thoroughly, with a steady stream of commentary from the *Amplifier* and a set of carefully-thought-out resolutions buttressed with arguments in their support. In order to have a command post where the supporters of change could meet and perhaps convert the doubters, RFMDA secured space in the Henrose Hotel. They invited John Burke to visit their headquarters, but Burke declined the invitation.[1]

The Rank and File Movement came to the convention with a handbook for constitutional reform and a call upon the delegates to rise to greatness. The September issue of the *Amplifier* commented:

> The delegates to our Detroit convention are called to Greatness. Our times and our Union's position leave no middle ground. You delegates will be *Great* or you will be *failures.*
>
> This convention is different and every delegate has to face up to that fact. The Trade Union movement is in serious trouble and every delegate knows it. Our Union's future is at stake and every delegate knows that too. Knowing it is not enough. You must *Act* or you betray those you represent.
>
> There is no "Easy Way" at this convention. The issues in plain language are:
>
> 1. Government by the consent of the governed.
> 2. The right of the taxpayer to know how his money is spent.
> 3. Proper conduct on the part of officers and representatives.
> 4. Justice for Union employees.
> 5. The BEST representation in bargaining at all levels.
>
> As we said, you delegates will be *Great* or you will be *Failures.* It's up to you. Never forget that generations of Unionists will be judging you on what you do because they will be affected by your decision. Don't kid yourself that this is not true. You are making history one way or another. Only YOU can decide whether you have the courage to make the right kind of history.[2]

RFMDA did not rely on rhetoric alone in its efforts to secure changes in the union. A booklet, entitled *Handbook for Constitutional Change,* was prepared which set forth the RFMDA position on the major points to be raised at the convention. The table of contents listed the general subjects of "Regional Elections," "Officers," "Per Capita and Expenditures," "Collusion and Corruption," and "Other Subjects" as the concern of RFMDA.

The main emphasis in the RFMDA drive for reform was to secure regional election of vice-presidents. In their *Handbook,* RFMDA set forth in detail six main points which summed up the reasons for regional election that had been advanced by the *Amplifier* and by the newsletters. The theme of the argument for regional elections was that this method

---

1. Burt Wells to John Burke, August 28, 1962, and Burke to Wells, September 1, 1962. Papers of the International Brotherhood of Pulp, Sulphite and Paper Mill Workers, Rank and File Movement for Democratic Action, Wisconsin State Historical Society, Madison, Wis.
2. *Amplifier,* September 1962, p. 1.

of selecting officers would make them more responsive to the will of the membership. The *Handbook* noted:

1. *The Union has regions now, but without representation.* Vice-Presidents are running the regions in which they work, but they are not responsible to the people whom they represent in talking with management or in making decisions on the Executive Board. When they represent *all* members like they say; they really represent no members.
2. *The Executive Board is now self-perpetuating.* Too many members of the Board have been selected not by the members but by the Board itself. No more than seven of the present 13 on the Board were originally selected by a Convention. They were named by the other members, and once named they naturally were elected at the next Convention because of the way elections are held.
3. *Vice-Presidents should represent the people they serve.* In the United States and Canada, the people elect their legislators on a regional basis, because this is the only way that a representative democracy can really function. This is the way Congressmen and Senators (are) elected in the United States and the way members of Parliament are elected in Canada. It works in government and it would work in the Union.
4. *Regional elections is the only way the members in a region can change their Vice-President if they are dissatisfied with the way he is operating.* Under present arrangements everyone in a region could be dissatisfied with the Vice-President, but the Vice-Presidents in other regions will support him at the Convention, and the delegates from the other regions will put him back in office.
5. *Unity is preserved by the Constitution.* Opponents of regional elections have said such elections would lead to disunity. But there is no evidence of this in our governments. There are different views and different interests in the different areas, and these are reconciled in the common interests of the people through the laws that are passed and enforced. It is the same in the Union. The differences between the different areas are reconciled in the Constitution, which is the law of the Union and must be enforced by all the Officers.
6. Opponents also say that there would be too much politics under regional elections. Other unions have regional elections and get their work done in good order. We have had more politics by International people this past year than we could have any other way.[3]

The *Handbook* contained a set of alternative resolutions for the delegates use on the convention floor and a detailed plan for dividing the continent into twelve regions. The proposed regions had been set out by Melvin Melton of Local 194, Bellingham, Washington. The plan was for the regions to correspond roughly to the areas already informally established, but with boundaries defined for the first time. Melton had set out regions that not only listed states, but divided some states, with counties listed according to the regions to which they would belong.

3. Rank and File Movement for Democratic Action, *Handbook for Constitutional Reform*, August 1962, p. 7. In possession of the author.

This was a proposal the delegates could understand, not a mere declaration of principle. If this resolution was to be adopted, there would be clear-cut boundaries setting off one area from another. Under the proposal, the lake states region of Wisconsin, Minnesota, and Upper Michigan would have been the largest in terms of membership with 21,331 members, based on per capita tax payment for 1961. The smallest region would have been the western states of Arizona and California, with total membership of 5,745.

The handbook went on to spell out in some detail the RFMDA position on such issues as compulsory retirement of officers, separation of the offices of president and secretary, opposition to an increase in the per capita tax, presentation of charges of corruption and collusion, and support for the resolutions favoring the reinstatement of George Brooks and Paul Hayes.

Thus, the RFMDA came to the convention floor in Detroit with a plan for change, and the mechanism through which that change was to be achieved. It became apparent soon after the start of the convention that RFMDA would be fortunate to win any of its reforms, let alone the bulk of its program.

## *The Detroit Convention*

Before the convention sessions had formally opened the RFMDA suffered a reverse when two of its staunch supporters, Angus Macphee of Local 708, and Orville Braaten of Local 433, both in British Columbia, were detained at the border. Braaten apparently had some feeling that the possibility of detention at the border was likely, since he wrote to Burt Wells in February 1962 that "It could be well that in the Detroit Convention some Canadian delegates may be stopped at the border."[4] President Burke appealed to Attorney General Robert Kennedy to investigate the situation and let Macphee and Braaten into the country to attend the convention. Burke's telegram to Kennedy concluded:

> These men are not dangerous to the safety or security of the United States. I know them personally. They are not Communists. It is true that they are somewhat radical in some of their views, but surely the United States of America, which was born in revolution, is not afraid of radical ideas. We need some radical delegates at our Convention to stir things up and make it interesting.[5]

Although the Attorney General did not heed Burke's plea, the convention was certainly stirred up and interesting, even without the attendance of the two radical delegates.

4. Papers, Orville Braaten to Burt Wells, February 6, 1962.
5. International Brotherhood of Pulp, Sulphite and Paper Mill Workers, *Proceedings of the Twenty-Sixth Convention,* September 10–16, 1962, p. 405.

The pattern the convention was to follow emerged in the session on the afternoon of the first day, September 10. Resolution No. 1, proposing amendment of the constitution by inserting a section setting forth the objectives of the union as protecting and strengthening democratic institutions, aiding the unorganized, assisting the AFL-CIO, protecting the labor movement from collusion and corruption, and safeguarding the democratic character of the labor movement, was introduced by Local 68, Oregon City, Oregon, a supporter of RFMDA. The Resolutions Committee recommended nonconcurrence. Delegates Perrin and Melvin Melton, among others, spoke for the resolution. William Perrin noted:

> I do not believe in good conscience, people can vote against a resolution of this nature, to set our aims and objectives before the whole world to believe in.
>
> It is approved language. It is good language. There is no particular reason why anyone should have any fear of it.[6]

Vice-President Louis Lorrain, speaking against the resolution, noted that its proposals were already covered in the union constitution and that the Constitution of the AFL-CIO dwelled upon the subject at length. Since the Pulp Workers were affiliated with the AFL-CIO, Lorrain felt they were covered by the provisions of its constitution and did not have to change their own. The resolution was defeated.[7]

Some resolutions favored by RFMDA went down to defeat without much of a fight being put up. This was the case when resolutions favoring election of officers by secret ballot and the proposal for separation of the office of president-secretary were introduced.[8] The RFMDA did fight vigorously, however, for their resolution proposing establishment of a new office, an executive vice-president. Melvin Melton and other RFMDA supporters spoke in behalf of the resolution. Melton cited figures for membership in some large cities around the country and said the relatively low membership showed the need for an executive vice-president to supervise organizing efforts.[9] Opponents of the resolution stated it would create a super-vice-president who could come into the territory serviced by another and tell him what to do. The minority report supporting the resolution was defeated.[10] The next resolution of interest to RFMDA was concerned with establishing a controller to supervise the accounts of the International Union. The proposal directed that the controller employ a certified public accountant to audit the books of the union. The report of the committee recommended non-

6. *Ibid.*, p. 22.
7. *Ibid.*, p. 25.
8. *Ibid.*, pp. 26–27.
9. *Ibid.*, pp. 27–28.
10. *Ibid.*, p. 38.

concurrence and RFMDA went into action. Melvin Melton stated that the rationale behind the resolution was to broaden the duties of the International auditor and make him a controller. The auditor's job was part-time, taking about six weeks a year. The proposal for a controller looked to a full-time job supervising the accounts of the union. Melton concluded:

> I submit this is something you should want, every one of you on the International Executive Board and every Representative in this Organization. You should want something like this. You should want it so you do not have to constantly be answering questions, defending your expenditures, defending every dollar you spend. You people should want this. I cannot understand why you are rejecting everything here.
>
> I cannot for the life of me understand how you can reject every positive suggestion that is put forth.[11]

Treasurer Henry Segal, who had been auditor from 1947 to 1960, spoke against the proposed constitutional change. It was Segal's feeling that the financial reports of the union were models of trade union accounting. He commented:

> I think we have an excellent report. I have made analyses of the reports of International Unions throughout the entire North American Continent, and I challenge anyone to show where any International Union gives as complete and comprehensive an outline of expenditures and receipts as our International Union does as far as the Auditor's report is concerned. In addition to that, no union goes into the detail our Organization does as far as listing the expenses and receipts.
>
> Furthermore, the quarterly reports that are issued and given to our membership certainly gives our people a complete outline and comprehensive view as to the operation in respect to your money you have entrusted in your President-Secretary and in your Treasurer.
>
> I think a job has been done as far as the finances are concerned. If there is ever a watchdog, that watchdog is John P. Burke.[12]

Most of the other arguments on the issue were variations of these points of view. Proponents of the controller felt the organization had outgrown its system of supervising its accounts, while opponents felt the system was adequate for the job. When the vote was taken, RFMDA lost again. The committee report recommending nonconcurrence was accepted, 633 to 355.[13]

With RFMDA sustaining defeat regularly on the substantive aspects of its program, it was apparent that the reform group lacked the strength to secure the changes it sought. In the conduct of debate on the various resolutions, President Burke was very liberal in permitting a free exchange

11. *Ibid.*, pp. 47–48.
12. *Ibid.*, p. 56.
13. *Ibid.*, p. 60.

of views. While there were some complaints after the convention that reform supporters were not recognized when they attempted to speak, there does not appear to have been any pattern of discrimination between the opposing factions. President Burke could have easily recognized only administration supporters, since he well knew the backers of RFMDA. He did not discriminate, and the fight went on, reaching a climax in the central issue raised by the reformers, that of regional election of vice-presidents.

## *The Regional Election Debate*

The question of regional election of officers was an issue of paramount importance to the reform group, and the debate on this proposal was the high-water mark of the reform effort at the convention. Some idea of the intensity of feeling on the issue can be gained from the *Proceedings*. There were twenty-seven resolutions on the subject introduced, and the debate occupied the convention for most of the afternoon of September 11, continuing into the evening and past midnight.

The majority report on the regional election question recommended nonconcurrence with the resolution. A minority report supporting the resolution was read and the debate began. William Perrin opened the discussion. He stated:

> On area-wide Vice-Presidents we feel this is a workable resolution. It offers the opportunity to agree in principle and to work out a program that would be satisfactory, I am sure, to the people assembled at this Convention.
>
> It is a sound system that protects each area by a showing of assurance, by assuring, I should say, a Vice-President who is the choice of the people in the area.[14]

Support for the concept of regional elections was not confined to the West Coast delegates. Carl Kaehn, from Local 52, Port Edwards, Wisconsin, supported the resolution.[15] Another Wisconsin delegate, Rufin J. Skibba of Local 482, Neenah commented:

> I do believe President Burke, that all of our Vice-Presidents are servicing us in a manner we, as members of this International Union want to be served. But I do believe the selection should be made by the people who know and directly have contact with the man who represents them.[16]

Not all the speakers rose to favor the regional election proposal, for there were those who objected to the idea. Delegate L. A. Crosby of Local 435, Savannah, Georgia, noted:

---

14. *Ibid.*, p. 74.
15. *Ibid.*, p. 76.
16. *Ibid.*, p. 77.

> I rise against the minority report because I was on the Committee on Constitution, and I think that every International Officer should come to this Convention and face every one of us, not merely the people he is going to represent in his area. After all, he is representing the International Brotherhood of Pulp, Sulphite and Paper Mill Workers. If he is going to represent just the people in his area, just so he can be elected again, and then go and do as he pleases, there is nothing we at the Convention can do about this.[17]

Delegate Eugene A. Hagen from Local 100, Camas, Washington, a local that consistently supported the International Executive Board made a spirited defense of the committee decision to nonconcur in the resolution.

> I am a member of the Committee on Constitution, and I rise in defense of the Committee's action in recommending nonconcurrence in the resolution. I know it is going to be impossible for me to point out everything the committee considered in coming up with this recommendation of nonconcurrence. We thrashed it around for hours. I may also bring out some of the things that have been said already by some delegates. But I think it is my obligation as a member of the committee to point out some of the pitfalls of this position.
>
> We have heard people say this would not be giving the powers of dictatorship to a Vice-President. But I would like to point out one area where there is approximately 65 or 70 local unions. The per capita tax load in that particular area of the Vice-President would be controlled by four or five locals—that is, the per capita tax vote.
>
> I have always felt and will continue to feel a Vice-President is a representative on the International Executive Board of all the people of this International Union, not just a small minority group.
>
> Another example I would like to give you is this. A local union with about 100 votes would say to a Vice-President, "I would like to have you down here tomorrow; we have a guy that was laid off, Joe Doakes, so come on down." Then a local with 500 votes or so would call up that Vice-President and say, "We have six people who were discharged. Come on down."
>
> Well, the first local with 100 members could infer, "Listen Mr. Vice-President, if you want to get reelected in this region, you had better service us."[18]

Delegate William Walker from Sitka, Alaska, also spoke against the resolution and made what may have been the most powerful argument against the regional election of vice-presidents concept. Said Walker:

> And when I feel I am paying per capita tax into this International Union, I am a member, and I am entitled to vote on every Vice-President we have in this International Union.[19]

The staff of the International Union took little part in the debate, perhaps because they had assessed the sentiment of the delegates and

17. *Ibid.*, pp. 77–78.
18. *Ibid.*, p. 80.
19. *Ibid.*, p. 101.

found it to be against regional elections. During the debate, the *Proceedings* indicate that President Burke, in his role as chairman, allowed the proponents of the resolution ample time to state their reasons for favoring the proposal. In view of the long campaign that had been carried on in behalf of the regional election proposal, it is doubtful that the discussion won many converts. The decision on the resolution was made on the basis of a role call vote with three tellers. President Burke appointed Melvin Melton as one teller, the Canadian research director, Wilf Ostling, was another, and Wilbur Williams of Local 4, Palmer, New York, a supporter of the executive board was the third. The results of the vote indicated a sound defeat for the regional election proposal. It was voted down by a vote of 10,851 to 4,542.[20] The International Executive Board presented a solid front against the proposal, with one exception. Vice-President Godfrey Ruddick abstained from voting. Two International representatives, including John Eyer from the West Coast, voted in favor of the proposal. The defeat represented a rejection of the most important point in the platform of the RFMDA. It represented the defeat of the central issue that RFMDA had counted on to change the power structure of the International Union. The International Executive Board was fairly certain it had the votes to defeat the regional election plan. President Burke allowed more than enough time for full debate and despite the intensive campaign that had been waged for regional elections, they were handily rejected. There was to be no compromise with the advocates of reform.

RFMDA had introduced several other motions in its attempt to democratize the International. One was concerned with establishing recall provisions for elected officers. The committee recommended nonconcurrence and the resolution went down to defeat without any debate.[21] There was some debate, however, on the resolution introduced by Oregon City Local 68 proposing compulsory retirement of International officers, representatives, and employees at age sixty-five. The resolution excepted President Burke. The Constitution Committee recommended nonconcurrence. Vice-President Lorrain commented in explaining the committee action:

> We do not agree with the basic philosophy of compulsory retirement. We believe that concurrence in these resolutions will cause the loss of valuable experience of people now serving and doing good work for the organization who are above the age of 65.[22]

Delegate William Clute, Jr., from Palmer, New York, was one of those who presented the argument in favor of the principle of compulsory retirement. He said:

---

20. *Ibid.*, p. 121.
21. *Ibid.*, p. 142.
22. *Ibid.*, p. 148.

> But all of the people, to my knowledge, at the front of that platform have sat there and pushed retirement for the people back in the mills. We did not take into consideration how good a job they had done. We did not consider how bad a job they had done. We were not speaking of their personal record. We were talking about a principle.
>
> I think the thing you have here today is a little disgusting. You are all trying to stand up there and make martyrs out of yourselves. You are saying, "I'm 65 years old, and you are throwing me out the door."
>
> But that is not what we are doing. It is the feeling of the delegates that if it is good enough for the people back in the mill, why shouldn't it be good enough for you.[23]

Once again RFMDA was defeated, and it was clear that the reform forces were being frustrated at every turn, while the executive board position prevailed constantly. Evidently the board possessed sufficient strength among the delegates to carry the field when contests arose. The support for RFMDA was not as strong as had been anticipated, and although there were other issues to come before the convention, the pattern was clear: the International Executive Board was in control.

Other measures favored by the RFMDA that met defeat included an increase in the per capita tax and the establishment of a public appeals board. RFMDA questioned the need for an increase in the per capita tax and the propriety of certain expenditures. Once again the position of the International Executive Board prevailed. The same was true on the proposal to establish a public appeal board. Perhaps stimulated by the discharges of Representatives Connolly and Hayes and Research Director Brooks, the resolution, introduced by Local 400, Brunswick, Georgia read:

> Resolved, That a board of appeal be set up for the International Officers, Representatives and employees of this International Union.[24]

Arguments for the resolution made the analogy that the appeals board would be like the grievance machinery in the plant. Opponents of the resolution, including two International representatives stated they preferred to place their faith in the wisdom and justice of the executive board. Cash Price, a representative from the West Coast commented:

> I urge that if you are concerned about me and if you are concerned about protecting me with this appeals board, forget it.[25]

Representative Floyd Van Deusen opposed the measure in somewhat more detail:

> I would be affected by the board they are asking to establish. Why we have to be protected, I do not know.

---

23. *Ibid.*, p. 157.
24. *Ibid.*, p. 208.
25. *Ibid.*, p. 210.

> We have an Executive Board which I feel is honest and whose integrity is above reproach. Everyone of them is fair and honest. They have the best interests of this Organization at heart.[26]

Delegate C. L. Hughes of Local 947, Arkadelphia, Arkansas, had the final word in the debate:

> I thought I had seen some brown noses back in the plant, but I have never seen as many men trying to hold their jobs as I have in this Convention today.[27]

Once again, the reform proposal was defeated. It would seem logical for the staff of the union to have supported the concept of an appeals board. In the interval between the conventions of 1959 and 1962, three employees of the union had been fired under circumstances that were at least questionable. With that record in mind, other employees might have been expected to seek a greater degree of job security which might have been attained by passing the resolution. Evidently the representatives felt this was not the case. Apparently job security was to be obtained by supporting the administration position down the line, not by instituting a system of checks on the power of the executive board. Evidently, a system of personal authority prevailed, rather than the impersonal relationships and web of rules that characterize a more highly-bureaucratized institution. At least the behavior of the staff in opposing a suggestion in their own interest might indicate such a view.

## *The Effort to Reinstate Paul Hayes*

Frustrated at every turn, the RFMDA came to an item it felt was concerned with securing justice for a union member whom they felt had been unfairly discharged from employment. Paul Hayes had been severed from the International payroll because of his alleged political activities, mainly the circulation of charges against Vice-President Tonelli among the membership of the union. The resolution favoring Hayes's reinstatement was introduced by Local 375, Philadelphia, Pennsylvania. It read in part:

> Whereas, a majority of the Executive Board, ignoring the views of President Burke and acting in a manner contrary to all trade union principles, did unjustly and unfairly discharge Brother Hayes; therefore be it Resolved, That Paul J. Hayes be reinstated as an International Representative without loss of pay.[28]

The Committee on Resolutions recommended nonconcurrence and the debate turned on the minority report which favored the resolution.

---

26. *Ibid.*, p. 211.
27. *Ibid.*, p. 212.
28. *Ibid.*, p. 220.

In view of the importance of the question, the rule limiting the time that an individual speaker could take was waived, and Paul Hayes, armed with exhibits and records took the floor. After a recital of his background and trade union activities, Hayes began to get to the heart of his tale. He read from a letter written by Vice-President Tonelli to Francis Tierney in November 1958. The letter had reference to a labor agreement between the Albany Corrugated Container Corporation, a subsidiary of the St. Regis Paper Company and the United Steelworkers. The letter read in part:

> You will notice in the agreement that they [wage rates] are exceptionally high for the industry. I do not think these rates should be publicized in other corrugated locals we have under contract.[29]

The letter continued that the rates were so far out of line that the firm was thinking of shutting down because of its uncompetitive situation. Hayes stated that it was his feeling that even if another union made the Pulp Workers look bad, the employees of the membership had an obligation to give all facts and information, including information that showed the Pulp Workers lagging behind other organizations.

The former International representative recounted other instances of friction with Vice-President Tonelli. According to Hayes, Tonelli once instructed him to "put on a show" for the workers but not to quote rates or factual information. Hayes refused and quoted rates to the management in the negotiating sessions. In regard to the situation in New York City, Hayes stated that he was told that New York was a jungle, and different from other places. Hayes stated that he received his information on the Kempton article "The Dowry" and other activities that Tonelli was alleged to have participated in from International Representative Raymond Leon. Hayes reiterated that no charges were ever filed against him or Pat Connolly, that they were discharged over the protests of President Burke. According to Hayes, he was prevented from obtaining employment in the paper industry because Francis Tierney spread the word that the union would not look upon his employment with favor.[30]

In all of Hayes's presentation to the convention it is impossible to find an instance of a charge that was backed by solid evidence. As a result of information Hayes supplied to Senator Winston L. Prouty, the Bureau of Labor-Management Reports undertook an investigation of the Pulp, Sulphite Workers for possible violation of the Landrum-Griffin Act. After its inquiry, the Bureau could find no fault with the records or affairs of the union and gave them a clean bill of health. Hayes used correspondence between himself and Senator Prouty and between the

29. *Ibid.*, p. 225.
30. *Ibid.*, p. 236.

Bureau and the Senator to cast doubt upon the credibility of the statements made by the International officers and, in effect, show guilt by association. There was evidently some thought that since the Bureau was investigating, there was some wrongdoing. While there was preliminary indication of a violation of the law, the Bureau was ultimately to conclude that there were no grounds for filing charges against the union, and the investigation was dropped.

Throughout Hayes's entire talk before the convention, not a single piece of evidence was introduced which would indicate that the allegations made against Vice-President Tonelli had substance. With rumors and allegations about Tonelli being circulated around the convention, President Burke allowed the third vice-president to take the floor in his own defense. After a recital of his record in the trade union movement, he began to deal with the issues that had made his name a subject for controversy. Before getting down to the specific allegations that were being made within the union, Tonelli made this general statement:

> It has been said time and time again all over these United States, by Paul Hayes and one or two others, that there is wide-spread corruption in New York City. Where, I ask you, is this corruption in New York City or in the metropolitan area? Has one shred of evidence ever been produced to show that there is corruption in New York City? No. Has any local of our International Union ever been brought before the McClellan Committee? No. Where is this corruption? We defy anyone to show one single shred of evidence that there is corruption in this Union in the metropolitan area of New York or anywhere else in the United States or Canada.[31]

With this preliminary statement out of the way, Tonelli began to get to the specific questions that had been raised about his conduct. The first subject touched upon by the vice-president was the activities of Anthony Barbaccia. According to Tonelli's account, Barbaccia first became involved with the Pulp Workers Union in 1949 during a campaign to organize the artificial flower workers in New York City. Barbaccia was a leader in the campaign and assisted in a strike which resulted in a settlement at a firm which then employed about 500 workers. Barbaccia continued his organizing activity in the artificial flower industry and succeeded in enrolling about six or seven hundred workers. After a time, according to the Tonelli account, Barbaccia began to suffer from a series of heart attacks. Both Tonelli and President Burke advised him to retire in the interests of health. Tonelli wanted to recommend to Local 679 that it retire Barbaccia as its president on a pension of $75 per week but, according to Tonelli's statement, President Burke wanted the local to give Barbaccia a weekly pension of $100. At the meeting of the local where the retirement ques-

31. *Ibid.*, p. 249.

tion was to be discussed, Barbaccia called the meeting to order and then collapsed with another heart attack. While he was taken to the hospital, the local unanimously voted the $100 per week pension. Tonelli stated emphatically that the International did not finance the pension, the costs were absorbed by the local.

In regard to the statements made by the Union Research Director John McNiff to the McClellan Committee in 1957 that Local 679 had negotiated substandard contracts, Tonelli said:

> But substandard contracts from what point?
>
> If you compare the artificial flower industry with the pulp and paper industry or with the automobile industry and such industries, certainly there would be a difference from the standpoint of standards.[32]

Tonelli reported the results of the various investigations that had been conducted into his activities, character and associations. According to Tonelli:

> Paul Hayes, by his innuendoes and allegations has tried to assassinate my character. He is responsible for the Executive Board appointing a committee to investigate me. He is responsible for the Federal Bureau of Investigation investigating me for seven solid months. He is responsible for the officials of the Bureau of Labor-Management Reports investigating me.
>
> What do these investigations disclose? I will tell you what they disclose. They disclose that Vice-President Tonelli is not guilty in the slightest of the allegations of Hayes and his associates.[33]

One of the allegations made against Vice-President Tonelli concerned his home in Yonkers, New York. There was a picture of the house circulated throughout the union which purported to show the building to be worth $100,000 and to be located in an exclusive neighborhood. According to Tonelli, he built the house from a model that cost $38,000 and had a mortgage on the building of $25,000. The acre and one-half lot was purchased in 1942 for $3,500.

Tonelli concluded his remarks and the debate on whether or not to reinstate Paul Hayes began. It soon developed into a discussion of who said what to whom at what meeting, and as the hour was becoming late there was some pressure for a vote. When the delegates were counted, Paul Hayes's discharge was sustained by a vote of 728 to 264.[34]

## *Election of Officers*

When the time arrived for election of officers, the fears of RFMDA that an area could be forced to take a man it did not want were con-

32. *Ibid.*, p. 251.
33. *Ibid.*, p. 253.
34. *Ibid.*, p. 276.

firmed. There were three contested offices, with RFMDA sympathizers or supporters involved in each one. Perhaps of greatest significance was the fact that a contest developed for the office of fifth vice-president. This was the job held by Godfrey Ruddick, who had opposed the executive board majority on the issues of the discharges of Connolly, Hayes, and Brooks. Ruddick was opposed in his bid for reelection by International Representative Reeves H. Brunk. When Ruddick was nominated, a petition supporting his reelection was read into the record that expressed support for his candidacy. The document was signed by 86 of the 89 delegates present at the convention from the area serviced by Ruddick. The petition noted that the delegates were "completely satisfied with the way he [Ruddick] has represented us during the past three years."[35] With an expression of support such as that from the people served, it might have been expected that Ruddick would have been in a good position to win reelection handily. This did not prove to be the case, for the vote was very close. When the executive board was polled it became apparent that Brunk was their candidate. Ruddick received two votes from his colleagues on the board. President Burke voted for him as did International Auditor Keith Wentz. Ruddick voted for himself. The other members of the board voted for Brunk, with the exception of four members who abstained. The final tally showed Ruddick reelected by a narrow margin of 8,004 to 7,202 for Brunk.[36] This in spite of the fact that he was clearly the choice of the people he was to represent.

There were two contested elections in the areas closer to RFMDA's strength, the West Coast of the United States and Canada. When the office of ninth vice-president was open, the incumbent, Oren Parker was renominated, and International Representative John Eyer opposed him. Twenty of the thirty-six local unions in the region served by Parker supported Eyer. The twenty unions that supported Eyer's candidacy represented about two-thirds of the membership in the region. After a short discussion, the role was called and Eyer was badly defeated, losing 11,021 to 3,613. He received two votes from the executive board, Godfrey Ruddick and Keith Wentz cast their ballots in his favor.[37]

There was one other contested position, the vice-presidency of the Western Canadian region. The incumbent, H. L. Hansen was renominated, and a supporter of RFMDA, R. B. McCormick of Local 312 was nominated to oppose him. Delegate R. Bryce of Powell River Local 76, who nominated McCormick, noted:

> It should be evident to the delegates assembled here in Convention that it is not possible to elect a Vice-President in this Convention that

35. *Ibid.*, p. 289.
36. *Ibid.*, p. 304. Ruddick was living on borrowed time. At the 1965 Convention of the Pulp Workers, he was defeated in his bid for reelection.
37. *Ibid.*, p. 324.

> is wholly suitable to the delegates of the area that the Vice-President will represent. It is, rather, the power and authority of the International Executive Board to pass around the word that they wish a certain Vice-President and then it is possible to elect that Vice-President.
>
> I think all of these things are only evidence that we made a mistake in turning down the area representation earlier in the Convention.[38]

Hansen was reelected by the overwhelming margin of 11,474 to 2,196.[39]

It was quite clear that the executive board was in complete control of the convention. Even where it was demonstrated that a vice-president was not the choice of the people he was to serve, he was returned to office. The worst fears of the RFMDA were confirmed at the convention. The executive board could afford to allow debate and contests for positions, secure in the knowledge they had the power to defeat the reform proposals. There were just too many delegates who heeded the word from the executive board for the reformers to succeed in achieving their objectives.

## *The Disposition of Other Planks in the RFMDA Platform*

The issue which had stimulated the formation of the Rank and File Movement for Democratic Action was the discharge of George Brooks from his position as research and education director for the union. When RFMDA came to the convention Brooks's reinstatement was still on their list of goals, though its priority had been reduced. When the resolution proposing the reinstatement of Brooks was introduced on the convention floor, nonconcurrence was recommended by the Resolutions Committee. Rather than open the floor to debate which might have opened a "Pandora's Box," Representative John Eyer moved the previous question on the majority report. The vote on the committee's recommendation favored acceptance, and there was no debate on the convention floor on the issue of restoring George Brooks to employment with the union.[40]

The resolution proposing restoration of Pat Connolly to the International staff was also disposed of without protracted debate. The committee advised against acceptance and this position was sustained. The remaining resolutions were disposed of in a flurry of activity, with the delegates following the recommendations of the various committees. The action of the delegates represented a rejection of the platform of the RFMDA.

The reform group had come to the convention with a set of proposals

---

38. *Ibid.*, pp. 324–325.
39. *Ibid.*, p. 340.
40. *Ibid.*, p. 352.

that had been developed in meetings and correspondence for more than two years. The ideas and concepts behind the RFMDA program had been well developed in the reform group's newspapers and in letters to the hierarchy of the union. There was no secret about what RFMDA sought or why it was seeking it. It is apparent, however, that in spite of the outrage over the ouster of Brooks, Connolly, and Hayes, over the alleged improper activities of Vice-President Tonelli, and over the supposedly undemocratic method of electing officers, the reformers never were in as strong a position as they thought. In effect, the advocates of change within the International were talking to themselves. Judging from the votes on the convention floor, in the campaign of over two years, RFMDA failed to gain many more supporters than it had when it was formed. The support it gained was balanced by the defections of such persons as George Lambertson and Mario Scarselletta, Jr. The reform group never had support in the large locals in the New York metropolitan area, and was forced to rely on smaller locals scattered around the continent. Judging from the votes, the appeal for change was not effective.

In part, the drive for reform may have been an effort by a younger set of personalities to assume some positions of responsibility and power within the union. John Burke had been president-secretary of the organization since 1917, and as long as he continued vigorous, there was no hope of displacing him. Nor was there any apparent desire to get rid of Burke. But there was apparent disquietude with the second-line leadership. With the exceptions of Vice-President Tonelli, the other men who were close to succeeding to the presidency were elderly. Given the age distribution of the officers, it was clear that Tonelli would some day be at the helm of the union, a prospect that some did not relish.

Judging from the actions of the leaders of the reform group, Perrin, Wells, Rogers, Melton, and Chatham, there can be little doubt that these people were sincere in their concern over an alleged lack of democracy, over charges of corruption and conflict of interest, and over the discharges of three International staff members. The leaders of the reform group lost many hours of work as well as their expenses in the drive for change. It is probably inaccurate to term the challenge to the executive board a straight power play. Had this been the case, there might have been a campaign to secure disaffiliation from the International, for contracts expired in the months following the convention. The RFMDA resolved to continue in existence, but in a meeting after the convention did not adopt a program.[41]

---

41. Papers. Circular letter, Burt Wells to RFMDA Area Executives, October 29, 1962.

## *The Aftermath of the Detroit Convention*

The immediate weeks and months after the Detroit Convention saw the RFMDA supporters discussing the events in a critical fashion. For example, the *East Coast Newsletter* assessed the convention as follows:

> We come to the conclusion as a result of what happened at Detroit, that it is impossible to make needed changes in this International Union. There are many reasons but the most important one is that President Burke and the Executive Board are in control of the machinery of the Union and are using it to prevent any change. This includes using the funds of the Union, the paid staff of Representatives and others and so on.[42]

Burt Wells reflected the situation of the Rank and File Movement in a letter to Peter Marshall that outlined the status of the reform movement and showed the thinking of some of the people on the West Coast.

> Needless to say I was quite crushed by the turn of events in Detroit and have no official comments regarding my feelings on the matter. I believe that the RFMDA should continue. I do not know how the movement is to be kept alive through merely publishing the *The Amplifier*. The George Brooks issue is dead as well as the Paul Hayes and Pat Connolly issues. Regional elections and other reform measures that were so vigorously supported by the RFMDA affiliates cannot keep the RFMDA in business. We took quite a beating at Detroit after two and one half years of preparation. What is the approach during the next three years?
>
> Progressive thinking union people here in the Northwest are troubled by two problems: One, they're very much concerned about the inflexibility of the Uniform Labor Agreement, and some of them are seeking means to get out from under its jurisdiction: Two, still other groups would go further and disaffiliate from the International if there were enough people behind that particular movement.[43]

While there was evidently intense disappointment with the results of the convention in Detroit, there does not appear to have been much sentiment for disaffiliation in the months immediately following—at least not in the United States. Three Canadian locals in British Columbia that had been affiliated with RFMDA broke away from the International and established the Pulp and Paper Workers of Canada. Angus Macphee and Orville Braaten were prominent leaders of the new union.

Apparently on the West Coast of the United States attention was being given to making the Uniform Labor Agreement more satisfactory to the people it covered. It was this change in emphasis that was to produce a split within the International.

---

42. *Ibid.*, "East Coast Newsletter," November 15, 1962.
43. *Ibid.*, Burt Wells to Peter Marshall, January 29, 1963.

## The Reformers Receive Advice

Some time after the close of the convention in Detroit, the reform group on the West Coast received a memorandum which analyzed the events at the convention and suggested a course of action to be followed. The source of the document is obscure, but it suggested a plan of action that was to be followed rather closely.[44]

The memorandum began by noting that reform of the International by the regular machinery was out of the question until 1965, when the next convention was to be held in New York City, Vice-President Tonelli's home ground. According to the memo, President Burke would not aid any reform movement and was trying to prevent change. While there were some kind words for the president-secretary, the author of the memo characterized his attitude as shallow, complacent, and self-satisfied. The alliance (if one existed) between Vice-President Tonelli, Office Manager Tierney, and Treasurer Henry Segal was termed a cabal, which was going to operate to frustrate challenges to the executive board.

The position of Vice-President Oren Parker was examined very closely. The memorandum noted:

> In spite of his personal incompetence, his position is strengthened in many respects. He will have the full-time efforts of the Representatives on the Coast, the official support of John Burke and the active assistance of those Vice-Presidents and employees who now regard Parker as one of their team; This adds up to a strong hand, *most of all because the Vice-President has the bargaining machinery so much in his hands.*

The writer of this document advised that the convention in 1965 was too far away to develop reform proposals immediately. It was advised that the reformers concentrate their efforts in the field of collective bargaining, especially on the West Coast. An alternative course of action was set forth as a last resort, "a move for independent unionism."[45] The document continued:

> Fortunately, the first steps toward an improvement in the local's position are the same, whether one is talking about reform internally or about independence. The local unions can therefore proceed with their work, as if the only possibility was to remain part of the organization. Having tried to do the job within this framework and succeeding or not, they can later make a separate decision about independence. Nothing is lost, whatever the ultimate objective.
>
> If independence proved to be necessary, this act would do as much for reform within the union as it would do for the integrity and posi-

44. The copy of this "Memorandum" in the RFMDA Papers in the Wisconsin State Historical Society contains an anonymous note indicating the memo was prepared by Sara Gamm, who was on the Research and Education staff of the Pulp Workers.

45. Papers, "Memorandum," p. 3.

> tion of the local unions that went independent. A successful move for independence on the part of a large unit like the Uniform Labor Agreement, joined possibly by other groups, would have a profound effect upon the internal workings of the Pulp and Paper Workers and the UPP. It would be a blow for freedom, within the Unions as well as for those who left the unions.[46]

The memorandum then went on to discuss the situation on the West Coast in some detail. According to the writer of this piece, since the convention forced upon the Coast a vice-president not of its own choosing, the Coast locals could go to extraordinary lengths to protect their integrity in contract negotiation and administration. According to this analyst, the Coast had leadership institutionalized in the Employees Association, a feature lacking in other areas of the nation. He also held that since the International had "usurped" much of the Association's functions over the years:

> A strong case can now be made that it is essential to the interests of the local unions and the members on the West Coast that the Association role be enlarged and made central in bargaining and contract administration.[47]

In other words, this was a proposal to reverse the tide of development that had prevailed in the preceding thirty years of collective bargaining relations between the employers and the unions. The proposal harkened back to the days when negotiations were less complex and the locals had a greater role in the process of forming and administering the contract, and it seemed to envisage a fragmentation of power and responsibility, in opposition to the movement towards consolidation of power that had prevailed in the preceding thirty years. Apparently the idea was to move against the great historical tendency discussed by Michels towards greater centralization of authority.[48] To be sure, the locals and their members have vital interests in the terms of the document they live under and the manner in which it is administered. Whether this proposal to enlarge the role of the Employees Association was the correct approach to the problem of securing more local autonomy is debatable. It is also a matter for question as to whether the Association was equipped to cope with the problems of negotiating and administering a complex document that had developed over thirty years, replete with interpretations and minutes.

After the general proposal to enlarge the activities of the Employees Association, the author of the memorandum went on to set forth a sug-

46. *Ibid.*
47. *Ibid.*, p. 4.
48. This concern with increasing the scope of local union activities has been voiced frequently by George Brooks, former Research and Education director of the Pulp Workers. Brooks has argued that the role of the local is central in preserving the sources of vitality in American trade unionism.

gested program for the West Coast. The program looked to a complete revision of the system that had developed since 1934. First, it was suggested that the Employees Association have the dominant and central role in negotiations and contract administration. This would have called for a full-time staff for the Association, in addition to machinery for getting the bargaining board established and the agenda formulated. The proposal also looked towards making the Association a party to the agreement, along with the International Unions. The next point suggested securing remission of some per capita tax to the Association so it would have resources with which to operate. It was also suggested that in the 1963 negotiations careful attention be given to the pension issue, with independent advice "so that the International Unions cannot thwart the desires of the local unions on this subject."[49] The final proposal advocated establishing a West Coast newspaper published by the Association.

With the proposal set forth, the author went on to develop supporting arguments for each point to serve as justification. On the scope of the Association's role under the proposal, the following was stated:

> Under the proposal the International Unions would continue to play a role, but negotiation of the ULA and its administration on a day-to-day basis would be the business of the Association, its officers and staff.[50]

That concept of the work of the Association does not leave much room for the International Unions. Though the author explicity denied that the proposal constituted dual unionism, it certainly appears to look to an organization other than the International Unions performing tasks traditionally carried on by the Internationals. As such, it seems to meet any test for dualism.

In regard to the remission of per capita, the author explained that since the Association would be performing tasks heretofore carried on by the Internationals it could expect a reduction in the amount the Internationals spent to service the Coast, and the Internationals could be expected to reimburse the Association for its expanded activities. It was also suggested that the West Coast have its own newspaper to cover the special interests in West Coast affairs such as arbitration decisions, since decisions became part of the ULA interpretations and binding on all member mills.

The document concluded with an explanation of the shift in emphasis from the RFMDA, national and international in scope, to a more limited concept of change, concentrating on the West Coast.

> The emphasis in this document is bargaining under the ULA. This emphasis is deliberate. It will be very difficult to maintain any lively

49. Papers, "Memorandum," p. 5.
50. *Ibid.*, pp. 5–6.

> program of national reform for the next couple of years. Since most of the really competent reform leadership is on the Coast, the prospect for a national RFMDA seems not very bright. It would seem better for the West Coast men to devote their energies and talents to winning autonomy for themselves. Their example would be an inspiration for others.[51]

In the course of setting out this proposal the author noted:

> John Burke and Paul Phillips [president of the UPP] will dislike the proposal, as will the field staff. Their principal weapons against the proposal will be that it is disruptive, would weaken the unions, would encourage dissension and possibly that the employers would not like it.[52]

These are formidable objections to the proposal. It is doubtful whether any chief executive wants to see his power and authority reduced. It is certainly true that the proposal was disruptive and would weaken the International, and to the extent employers seek to deal with an organization that can enforce discipline within its ranks, they probably would not like the proposal. (Unless they thought it would be easier to deal with the Employees Association.) Obviously the presidents of the Internationals and their staffs would be opposed to proposals that would reduce their power. This proposal looked towards reversing the tide of history in the collective bargaining relationship on the West Coast. It was bound to be resisted by those who had vested interests in the continuation of that relationship in the fashion in which it had evolved. The basic problem that this proposal was designed to solve was how to accommodate a contract covering a large unit to the problems of the local unions within that unit, and has been a problem of some importance in collective bargaining in recent years. The Steelworkers, the Auto Workers and the Electrical Workers have all been faced with similar problems. This proposal, however, looked to fragmentation of power and responsibility in collective bargaining, and was certainly an attempt to reverse the development of the relationship on the West Coast which had seen the Association cede power to the International in 1942.

As an outline for a course of action, the proposal was followed rather closely in the negotiations that followed its circulation. The Association did, indeed, attempt to increase its power at the expense of the International. When the International resisted, conflict developed, and it was this opposition of forces, the Internationals seeking to hold their position and the Association seeking to expand its influence that produced disruption in 1964.

51. *Ibid.*, p. 11.
52. *Ibid.*, p. 7.

# 7

# THE BREAKAWAY ON THE WEST COAST

## *Negotiations of 1963*

With the convention of 1962 past and the national campaign to reform the union defeated, the leaders of the reform element on the West Coast began to give thought to improving the system of negotiating and administering the Uniform Labor Agreement. Perhaps the anonymous memorandum discussed in the last chapter had some weight in stimulating thinking along this line, though as Burt Wells had noted, there already was concern about the Uniform Labor Agreement.[1] In the effort to modify the Uniform Labor Agreement, the anti-administration forces within the Pulp Workers were operating from a position of strength. They controlled most of the offices of the Employees Association and its subordinate bodies, and they had apparently gained control of a majority of the locals in the bargaining unit.[2] With a strong position in the Employees Association, the dissident group was able to control the negotiations for renewal of the ULA, since the Association played the main role in drafting the proposals to be submitted to the manufacturers.

The unions presented a lengthy agenda of fifteen items to the manufacturers in 1963 including a detailed proposal on pensions. Other items of significance in the union proposal were concerned with changes in the clauses relating to seniority, grievances and arbitration procedures. The employer agenda specifically granted the union request for bargaining on pensions. Before the negotiations began, however, the dissidents challenged the hierarchy on the matter of electing the chief union spokesman.

Under the structure of collective bargaining on the Coast, local delegates would assemble in Portland, Oregon, for the negotiations. They would arrive prior to the bargaining sessions in order to meet with one another and formulate the union agenda for the negotiations. This pre-

1. Burt Wells to Peter Marshall, January 29, 1963. Papers of the International Brotherhood of Pulp, Sulphite and Paper Mill Workers, Rank and File Movement for Democratic Action, Wisconsin State Historical Society, Madison, Wis.

2. Leonard Kattan, "The Private Settlement of Industrial Disputes," (Unpublished Paper, The Graduate School of Business Administration, University of California at Los Angeles, January 1966), p. 40.

liminary meeting was termed the prewage conference. In addition to formulating the union agenda, the delegates elected eight of their number to serve with the three West Coast vice-presidents and the union staff members as the bargaining board. The board conducted the negotiations under the eye of the remaining union delegates.

Until the negotiations of 1963, the chairman of the union bargaining board was elected by the board composed of eight local union delegates and the International officers and staff. Prior to 1959, election to this position had been a rather routine affair, since there was apparent agreement that John Sherman, vice-president of the Pulp Workers, would serve as chairman. With Sherman's retirement, and the merger of the Papermakers and the Paperworkers, the chairmanship evidently became a position open for contest between the UPP and the Pulp Workers. In 1959, Vice-President Ivor Isaacson of the Pulp Workers was challenged by Vice-President Al Brown of the UPP. Since the Pulp Worker delegates outnumbered those from the UPP by a considerable margin, Isaacson won handily. Following the contest in 1959 there was apparently some arrangement worked out between the two unions to put the chairmanship on a rotating basis. Thus, Al Brown was elected without opposition in 1960. In 1961 it was expected that Oren Parker of the Pulp Workers would be chairman, but Parker withdrew his candidacy, expressing his belief that he did not feel qualified to handle the conference. Brown was then elected conference chairman for the second year in a row. In 1962, Vice-President Parker of the Pulp Workers was elected chairman without opposition, but his conduct of the negotiations did not please some of the delegates. Thus, the position of chairmanship became a concern of the union delegates as the 1963 sessions commenced. In keeping with the rotation concept, the International proposed UPP Vice-President Oscar Robertson as union chairman for the 1963 negotiations. The delegates, however, evidently were not disposed to follow the wishes of the Internationals in regard to election of the union chairman, although they were not necessarily opposed to Robertson who never had chaired the sessions in the past. Thus, at the prewage conference of the delegates prior to the start of negotiations, the delegates adopted a resolution calling for election of the union chairman by the delegates from the local unions rather than by the bargaining board. The resolution continued by stating that the International vice-presidents would automatically be candidates, though further nominations were not excluded. The resolution carried by a vote of 92 to 40. When the time came for election of the chairman, Oscar Robertson of the UPP was elected with 77 votes to 60 for Ivor Isaacson of the Pulp Workers and 5 for Oren Parker, also of the Pulp Workers.

Having successfully changed one of the practices for the conduct of negotiations, the delegates began their sessions with the employers on their fifteen item agenda. Before the conference was very old the manufacturers made a statement regarding publicity and reports of the negotiations. Sid Grimes, of the Manufacturers Association stated:

> Another thing that we have agreed upon jointly here over the years is our policy on publicity. I am now talking about publicity while the negotiations are in progress and until they are concluded. The policy we've followed has been that no information whatever is to be given out or released about our Agendas or our deliberations here except to our official Publicity Committee. This means that no information is to be telephoned back to our mill communities as well as none to the press.[3]

Speaking for the unions, Oscar Robertson agreed with this interpretation.

With that statement in the record and the exchange of agendas completed the work of the conference began. The bulk of the negotiating time from May 3–24 was taken up with a discussion of the pension issue. Among the delegates there apparently was a feeling that the International officers came to the negotiations unprepared to discuss the subject. There was a feeling that:

> Going into the 1963 conference with the knowledge that pensions would be a subject for bargaining, the International unions were totally unprepared. It seemed to many that they were desirous of creating such confusion on this issue that the unions would finally yield to the original company position even though they had the Court's decision on their side.[4]

Although the Internationals may have been unprepared as charged by the locals, the delegates themselves had come to the negotiations with a detailed proposal to place on the table. Local 194 of Bellingham, Washington, Melvin Melton's home local, had prepared a detailed plan which was adopted by the prewage conference with only minor changes. The Bellingham proposal served as the basis for the union position on the matter of pensions.

At the same time that the unions came to the conference with a proposal on pensions, the manufacturers, too, came prepared with their own plan. After it was presented to the unions, Oscar Robertson, chairman of the union delegation commented that the union representatives had rejected the plan as completely unacceptable.

Some of the union delegates from the locals covered by the ULA expressed their reaction to the employers proposal on pensions. Fred Delaney, a member of the bargaining board commented:

---

3. International Brotherhood of Pulp, Sulphite and Paper Mill Workers and United Papermakers and Paperworkers, and the Pacific Coast Association of Pulp and Paper Manufacturers, *Record of Negotiations*, May 3–24, 1963, p. 38.

4. Papers, Affidavit of Burt D. Wells, State of Oregon, County of Multnomah, August 9, 1964, p. 8.

> From the opening rejection of the Unions' and the subsequent committee meeting, the Manufacturers are conveying to us their intent to rub our noses in the dirt because the Unions won the court decision forcing the Manufacturers to bargain on pensions here in Portland. If you want to promote this type of atmosphere to bargain in, go right ahead, this is another of Management's prerogatives.[5]

The main union objection was to that part of the manufacturers' proposal that stipulated that inclusion of a clause on pensions in the ULA "eliminated bargaining on pensions at the company level and also eliminated bargaining on company pension plans at any level. . . ."[6] The unions refused to renounce the right to bargain at the local level, although their proposal looked to the establishment of certain Association-wide standards in the area of pensions.

The negotiations in the area of pensions consumed the bulk of the time at the conference in Portland. Much of the time was spent in repetition of the various positions. On the afternoon of Tuesday, May 14, Oscar Robertson of the UPP suggested to the manufacturers that a joint committee be named to study the pension question. The proposal was acceptable to the manufacturers and the conference recessed to wait upon the deliberations of the joint pension committee. Four days later, on the afternoon of Saturday, May 18, the conference reconvened to hear the proposal that had been worked out. Oscar Robertson notified the employers that the union delegates were convinced the conference had made as much progress on pensions as was possible and that the delegates "reluctantly" agreed to adopt the report of the joint committee on pensions.[7] The solution reached on the impasse of local bargaining called for the companies to appoint responsible people with whom union representatives could meet for "discussion in good faith" on ideas or problems in the area of pensions.[8]

When the conference was concluded on May 24, the unions had made some progress towards meeting the objectives stated in their agenda at the start of the conference but were still short of their objectives. Wage rates had been increased 7½ cents per hour, half the union proposal of 15 cents per hour. The benefit plan had been liberalized, and the provisions in regard to adjustment of grievances had been modified to come closer to the membership wishes on the subject. In addition, the unions had secured language on pensions that conformed closely to their original proposal. The manufacturers concluded that the package they had agreed to would cost them 9.63 cents per hour, an increase, as they calculated it, of 2.98 per cent. After the final Manufacturers Association offer was

---

5. *Record of Negotiations*, p. 102.
6. *Ibid.*, p. 85.
7. *Ibid.*, p. 238.
8. *Ibid.*, p. 245.

spelled out the union delegates caucused, and Oscar Robertson reported the result:

> I have been instructed to inform you that these delegates cannot accept your final position. They have unanimously gone on record, and I say to you that they have gone on record to this extent, that they will refer your final position back to their membership with a recommendation for rejection.[9]

## *Rejection of the Uniform Labor Agreement*

Following the recommendation of their delegates, the membership voted to reject the contract terms negotiated in the sessions from May 3–24. The vote for rejection was overwhelming, 12,515 against acceptance to 2,745 for the contract.

The negotiations were reconvened on July 9, the earliest date the manufacturers could arrange to have the necessary personnel present. President Burke of the Pulp Workers sought an earlier start for the negotiations, but could not persuade the companies that prompt negotiations were desirable. At the start of the negotiations on July 9, the Manufacturers Association raised a point about publicity. Sid Grimes, the manufacturers' representative, asked Oscar Robertson if the arrangements concerning publicity surrounding the negotiations still were in effect. Robertson replied:

> As far as the press release is concerned, the delegates from the Unions have instructed me to inform you that the Unions—the Bargaining Board Chairman, I should say, which is me—will have the responsibility of determining whether or not there will be a joint press release or whether or not the Unions will release—make a press release—of their own. And so the responsibility is going to rest with me to make this determination upon the conclusion of this Conference.
>
> As far as the commitment that we have had in the past to maintain no contacts between the delegates and the members of our local unions, I have been instructed to inform you that we are no longer desirous of continuing this situation at this time. During this Conference the delegates reserve the right to report to their local unions at any time during these negotiations.[10]

As the negotiations proceeded the manufacturers from time to time referred to the union position on publicity and stated they wished it would be revised. Sid Grimes, the manufacturers' representative, noted that when the system of bargaining was adopted one of the rules agreed upon was the prohibition of management or union delegates releasing information to constituents or anyone else during the course of the negotiations. Grimes also noted:

9. *Ibid.*, p. 446.
10. *Record of Negotiations*, July 9–20, 1963, pp. 29–30.

> . . . This Conference is merely a continuation of the one we started in May and we do not think it is fair or proper to come in now and want to change the rules in the middle of the game.[11]

The Unions refused to reconsider their position on the matter of publicity as Grimes had requested. They reassured the manufacturers that they did not intend to post progress reports on bulletin boards but that responsible people in the local unions had a right to know what was taking place at the conference. Grimes noted that the manufacturers felt it was unfair to change the rules of the game once they had been agreed upon.[12] Robertson commented that the delegates were:

> . . . acting under instructions from their local unions and their local unions are footing the bill. As long as they are paying the bill, I guess they're going to have to do what they're told to do.[13]

The manufacturers were evidently very upset that the unions were modifying the agreement on publicity that had governed the earlier session of the conference. After the unions had refused once again to reconsider the matter, Sid Grimes commented:

> The Manufacturers want you to know, however, that they view and believe your position to be an improper exercise of unilateral action on your part in changing the procedural rules of this Conference which you agreed to while it is still in progress.[14]

Grimes's statement ended the discussion of the publicity issue for the remainder of the conference, but its implications were to return to bedevil the parties in the months ahead.

The conference concluded with the addition of a revised seniority clause that substantially met the requirements set forth by the union delegates. The result of the two sessions gave the unions seven of their original fifteen demands. When the amended result of the negotiations was submitted to the membership it was accepted.

The discussion on the issue of publicity may have represented a step in the direction of bringing the negotiations closer to the rank and file member, a course of action suggested in the memorandum received by the West Coast leadership shortly after the 1962 Convention. The emphasis on reporting to the locals the status of the talks represented a departure from the historic practice of withholding information until the close of the conference when a joint press release was issued. Reporting the progress of the talks back to the locals was in line with the idea of increasing the participation of the locals in the negotiations. This was a move suggested by the memorandum. As such, it might have represented a first step in the erosion of the position of the Internationals, a

11. *Ibid.*, p. 72.
12. *Ibid.*, p. 86.
13. *Ibid.*, p. 90.
14. *Ibid.*, p. 99.

goal of the author of the memo and likely an aim of the West Coast leaders. If the attempt to change the conduct of the negotiations had been unchallenged, the Westerners would have been emboldened, perhaps, to take further steps to whittle away the role of the International and increase the part played by the Employees Association. However, the intraunion contest for power operated to place the employers in a situation in which the rules of the game were shifting without their knowledge, direction, or consent. This was a situation intolerable to the companies covered by the Uniform Labor Agreement and they moved to assert some independence of their own. It was this move by the manufacturers that was to lead to schism within the unions.

## *The Ground Rule Debate*

The members of the Manufacturers Association were evidently quite upset over the unions' action to change the format of the negotiations during the conference, and they determined to take action to forestall such situations in future negotiating sessions. On December 20, 1963, the Manufacturers Association mailed a letter to President Burke of the Pulp Workers, President Phillips of the UPP, and the vice-presidents of the unions who had responsibility for the area covered by the Uniform Labor Agreement. In the letter, the manufacturers referred to the dispute over publicity in the 1963 negotiations and set forth a written codification of the rules to govern future negotiating sessions. The manufacturers wanted to make it very clear that they regarded the International Unions to be the legal bargaining agent for their employees. They stated that:

> The bargaining agent for "employees" as that word is defined in the Uniform Labor Agreement, has been and is now the . . . two International Unions, acting jointly as a single bargaining agent and hereinafter collectively referred to as the Internationals. . . .[15]

The Manufacturers continued by noting that as the bargaining system on the West Coast developed the Internationals and the Association informally adopted certain practices which came to be regarded as essential for satisfactory Association-wide negotiations. Prior to the negotiations in July 1963, neither party had sought any substantive revision in long-established negotiating practices. The manufacturers noted that one essential practice was the prohibition of publicity during the negotiations. They commented:

> Although the Association is sure that both it and the Internationals regarded the disclosure prohibition as a continuing agreement, it was restated and reconfirmed at the beginning of each negotiation confer-

15. Papers, Pacific Coast Association of Pulp and Paper Manufacturers to John Burke, Paul Phillips, and others, December 20, 1963.

> ence solely for the purposes of (1) bringing the disclosure prohibition to the attention of persons who had not attended prior negotiations and (2) refreshing the minds of all other persons attending the conference.[16]

The manufacturers made plain their feeling that the Internationals had not lived up to their responsibility as the legal bargaining agent. The employers felt that the change in the publicity rule came from the locals and thus represented a usurpation of International power and responsibility. The letter noted:

> At the opening of the reconvened conference on July 9, 1963, the Union co-chairman astonished the Association by announcing the unilateral abrogation by the Internationals of the disclosure prohibition. He made it very clear that the Internationals were directed to take such action by the delegates from the Local Unions. The Union co-chairman stated, in effect that many of the delegates had, prior to being sent to (the) reconvened 1963 negotiations, been ordered by the Locals which "sent them to Portland and paid the cost of keeping them there" to report to their constituents anything that transpired at the negotiations which any delegate might choose to report. In other words, as the Association views the incident, the Locals were being allowed (1) to usurp the rights and responsibility of the Internationals and (2) to enforce, in the middle of a negotiation conference, revisions, of a practice previously considered essential.[17]

In order to prevent a repetition of events such as those that had occurred in 1963, the manufacturers proposed that certain procedures to govern the conduct of negotiations be reduced to writing. Their proposal spelled out in some detail the ground rules they wished to see in effect at future negotiating sessions. The proposal envisaged two co-chairmen, the unions' chairman to be selected by the Internationals by whatever method they desired. The manufacturers proposed that only members of the bargaining board could participate in negotiations and that members of the union bargaining board would be "designated by the Internationals."[18] The manufacturers also stipulated that a non-disclosure rule would be read into the record of every conference. Other proposals made by the manufacturers were in the area of conducting a referendum on the employers' offer and the stressing that it was the duty of the Internationals to conduct the negotiating sessions. The manufacturers also stated that changes in ground rules could be proposed in writing at least 120 days prior to the appropriate June 1 prior to a negotiating session. Changes could never be made during a negotiating conference.[19]

Thus the manufacturers were proposing for the first time a written

16. *Ibid.*
17. *Ibid.*
18. *Ibid.*
19. *Ibid.*

codification of the rules to govern the negotiating sessions and at the same time they were evidently attempting to bolster the authority of the International Unions. Their proposal clearly looked for a diminution of the role of the locals and a proportionate increase in the authority of the International. In view of the ferment on the Coast, of which the manufacturers cannot have failed to be aware, their emphasis on the power of the Internationals at the same time that the locals were attempting to increase their own influence made some type of conflict almost inevitable. A situation existed where the the Internationals, with the support of the companies, were attempting to increase their authority. At the same time the locals were seeking an increased role in the bargaining and administration of the contract. These two conflicting forces, operating together, were to produce a collision.

## *Development of the Ground Rules Controversy*

The hierarchy of both the Pulp Workers and the UPP considered the ground rules proposed by the manufacturers and found them unacceptable. At the same time, however, they may have been desirous of asserting their authority over the local unions that were under the Uniform Labor Agreement. Thus, the idea of written ground rules for the conduct of the negotiations was not abandoned. The principle was accepted. After some consideration the presidents and vice-presidents of the two Internationals called a meeting for February 10, 1964, in San Francisco with the men who had been elected in 1963 as members of the union bargaining board. These local union members had no status in 1964, since the composition of the bargaining board is determined yearly at each conference, and the board is not a continuing body. The members of the 1963 board agreed, however, to meet with the officials from the Internationals though they had no official status in 1964. At the meeting on February 10, there was apparently no mention of ground rules for negotiations. The subject discussed concerned the procedures involved in taking strike action against the manufacturers. The membership of the 1963 bargaining board left the meeting feeling some progress on the question of strike had been made. At this time neither the local unions nor the hierarchy of the Employees Association nor the members of the 1963 bargaining board had any knowledge that the manufacturers had proposed reducing the procedures for conducting the negotiations to writing or that their International officers had accepted the principle. The next day, the officers of the Internationals who had just met with representatives of the locals on the West Coast went into conference with the manufacturers on the issue of ground rules. This was unknown to the locals until the International presidents informed them of the results of the conference ten days later. In a letter to all locals covered by the ULA the presidents of both Internationals, the UPP and the Pulp Workers, set out their policy toward multi-plant bargaining conferences and the procedures to be followed in conducting the 1964 negotations under the ULA.

The letter to the locals contained two documents, one was the ground rules for the conduct of the 1964 negotiations and the other was the text of a resolution the Pulp Workers Executive Board had adopted in their meeting of January 13–18, 1964, in Glens Falls, New York. The resolution of the executive board spelled out the position of the International Union on multiplant collective bargaining conferences. It was resolved that:

> 1. Local unions of this International Union whose members are affected by multi-plant collective bargaining contracts shall participate and be represented in such multi-plant collective bargaining conferences in a manner and to the extent determined by the Vice-Presidents or Representatives designated by the President-Secretary for that purpose.
> 2. Multi-plant collective bargaining conferences and union caucuses called in connection therewith shall be conducted in a manner, at a time and place, and in accordance with the procedures determined by the aforesaid International Vice-Presidents or Representatives.[20]

The ground rules adopted by the Internationals and the manufacturers for the conduct of the 1964 negotiations stipulated that Ivor Isaacson would be the chairman for the unions. If Isaacson was unable to serve for some reason, the union chairman would be appointed by the presidents of the UPP and the Pulp Workers, acting jointly. Other changes in the method of conducting the negotiations were spelled out by the ground rule agreement. According to the rules:

> The International Unions and the Association each shall choose its members of the Bargaining Board and [except by mutual consent] only members of the Bargaining Board shall participate in the discussions.[21]

The document on the conduct of negotiations also set out a strong clause on publicity. There was to be no disclosure of any information relating to the negotiations. The clause read:

> At the beginning of the negotiation conference, the Co-chairman for the Internationals and the Co-chairman for the Association shall each respectively state [to be recorded in the transcript of the conference] that any person attending the conference on his side will not, as long as the conference continues, disclose to any person not attending the

20. *Ibid.*, William H. Burnell and Paul L. Phillips to Officers and Members of ULA Local Unions, February 21, 1964.
21. *Ibid.*

> conference any information as to agendas presented and discussions and actions relating thereto in the negotiation conference.[22]

## *Action of the Employees Association*

Shortly after the meeting of the Internationals and the manufacturers on February 11, the Employees Association held its regular meeting in Portland, Oregon. The sessions lasted from February 19–21. The executive board of the Association met for two days prior to the general sessions and held one day open for a meeting with the officers of the Internationals, but the representatives of the Internationals did not make an appearance, and the meetings continued without them. At a meeting before the full sessions started, recommendations on ground rules were developed by the executive board of the Employees Association, which at that time was unaware of the agreement on ground rules reached between the Internationals and the manufacturers on February 11. On February 21, the same day the Internationals mailed the ground rules agreed upon in San Francisco to the locals, the executive board submitted its recommendations on ground rules to the delegates at the meeting of the Association. The delegates to the meeting overwhelmingly voted to adopt the recommendation of the executive board on ground rules. There were officers of the Internationals present at the sessions of the Employees Association, but during the discussion of the ground rules submitted by the officers of the Association, they did not make known to the delegates the fact that there had been a meeting between the Internationals and the manufacturers some ten days earlier. Nor were the delegates informed of the fact that an agreement had been reached between the Internationals and the employers on the rules for the conduct of the negotiating sessions which were due to begin in April. The International officials present at the convention of the Association made no objection to the adoption of rules for negotiation that differed from those agreed upon by the Internationals and the manufacturers.

The rules proposed by the executive of the Employees Association differed from those agreed upon by the manufacturers and the Internationals in two important respects. The Employees Association rules provided for election, not appointment of the union co-chairman of the bargaining board and the election of the local union representatives to the bargaining board. On publicity, the proposal provided that if lack of progress became apparent the delegates could return home after notifying the manufacturers of their intention to leave the conference. A referendum vote would then be conducted on the employers' offer.[23] These rules

22. *Ibid.*
23. *Ibid.*, Affidavit, p. 14.

were adopted by the Employees Association, but since it was the prewage conference that developed the agenda and discussed the employers counter proposals, it was felt it was up to the prewage conference to also adopt rules for the conduct of its negotiations. Since the officers of the Employees Association, William Perrin of Oregon City, president, Dick Archer of Everett, vice-president, and Elmer Lines of Los Angeles, secretary, were to become the officers of the prewage conference, the adoption of the Employees Association rules was facilitated.

## *The Prewage Conference and the Negotiations of 1964*

At the opening session of the prewage conference on April 15, 1964, Ivor Isaacson reviewed the developments that had occurred since the 1963 negotiations had concluded. He stated that the Internationals had refused to meet with the Employees Association in February because the International executive boards of the UPP and the Pulp Workers regarded the Association as a voluntary association of local unions representing members who were covered by many different labor agreements. The conditions under which a strike would be considered appropriate were outlined. Then Isaacson outlined the meetings between the International officers and the manufacturers held in San Francisco during February. He stated the agreement that was reached and sent to the locals represented the rules that would govern the conference. Isaacson noted that "any attempt by resolution or motion to change such rules for this year would be strictly out of order."[24] Delegate Graham Mercer complimented Isaacson on his presentation and moved that the conference elect its chairman. Isaacson replied that the motion was out of order, since the conference already had a chairman. Isaacson then went on to state that the credentials of the delegates were in order and that the conference was a duly constituted body. Melvin Melton objected to the actions of the chairman. He stated that the chair would be next appointing committees and the bargaining board. Isaacson replied: "You are absolutely right—we may do just that."[25]

Thus the tone of the conference was set. The International officers were attempting to squash opposition and run roughshod over the delegates who represented the membership covered by the contract the conference was called to negotiate. The International officers and staff were determined to conduct the conference according to the rules they had circulated to the locals, while the reform element which controlled

24. "Minutes" of the Pre-Wage Conference of the International Brotherhood of Pulp, Sulphite and Paper Mill Workers, and the United Papermakers and Paperworkers, Portland, Ore., April 1964, pp. 6–7.
25. *Ibid.*, p. 7.

the conference was attempting to conduct it according to the rules adopted by the Employees Association the preceding February.

The dissident element demonstrated the fact that it had firm control of the conference when officers were elected. William Perrin was elected chairman, Dick Archer, vice-chairman, and Elmer Lines, secretary. These were people who had supported the RFMDA and held executive positions in the Employees Association.

After the election of officers for the conference and the reading of the rules of order, delegate Fred Delaney moved that the delegates elect the conference chairman and that the International's vice-presidents, Isaacson and Parker from the Pulp Workers and Robertson from the UPP, be automatically nominated as candidates. At that point UPP Vice-President Robertson commented:

> You are simply spinning your wheels. As far as I am concerned Vice-President Isaacson is going to be your Chairman, and I will not stand for election this year.[26]

Delegate Moe Hunter commented in reply to the International vice-presidents:

> We simply do not like the dictatorial position the Internationals are taking and for this reason we will elect the Conference Chairman.[27]

The election was held and Isaacson was the choice of the delegates as chairman. He received 110 votes to 5 for Robertson and 4 for Parker.[28] Robertson stipulated for the record that the two Internationals considered that the election was incorrect.[29]

On Monday, April 20, the conference was called to order by Chairman Perrin who noted that the International officers had given the conference officers a set of resolutions regarding the rules for conduct of the negotiations. Perrin noted that the conference officers proposed rules to substitute for those advanced by the International officers. Both sets of rules were passed out to the delegates. Isaacson then stated that nothing in the ground rules prohibited the locals from putting out their own summary of the conference. He remarked:

> The International Presidents laid down the ground rules and we were told these are to be put into effect. The Manufacturers were also sent these same ground rules. Grimes will read these rules into the transcript and I will agree with them. We have our orders to abide by the ground rules irrespective of what you do.
>
> If you take any act on contrary to these rules, you are taking action that would not bc allowcd on this floor. You arbitrarily clcctcd mc even after Vice-President Robertson protested your action.

26. *Ibid.*, p. 10.
27. *Ibid.*
28. *Ibid.*
29. *Ibid.*

> I am asking you to abide with our ground rules. If I am Chairman, we will go by our rules.[30]

The rules adopted by the Employees Association in February were then passed out and the debate on their adoption began. The sentiment of the delegates as recorded in the minutes is overwhelmingly in favor of the ground rules introduced by Perrin and the other elected conference officers. Delegate Moe Hunter noted:

> After 30 years of successful bargaining, the International has taken this action. Last year the delegates made some minor changes and the result was a letter from Grimes to the Internationals. President Burnell[31] has never attended a conference and President Phillips has only been to one. Do not understand how these people in New York, not being familiar with the West Coast, can lay down rules we cannot live with. In my estimation, the Manufacturers laid down the ground rules. Both the Internationals and the Manufacturers are afraid of a strike on the West Coast. Think Grimes said if the Internationals did not agree to the ground rules there would be no conference. I have always been a firm advocate of the International Union, but these ground rules make me sick.[32]

On a roll call vote the rules of the officers of the conference were adopted 78 to 59.[33] At that point Ivor Isaacson made a statement:

> I told you the first day of this Pre-Conference that I was instructed to enforce the ground rules adopted by these two International Unions. Being an elected officer, I took an oath of office to uphold the Constitution and By-Laws and will do so. If I am the Conference Co-Chairman on Wednesday morning, the rules given us by these two International Unions are the rules I shall use.[34]

The negotiating session opened on Wednesday, April 22, and Sid Grimes, the representative of the Employers Association, began with a discussion of the ground rule problem. Grimes noted that in the middle of the 1963 negotiations the unions had unilaterally changed the rules for the conduct of the sessions and as a result the manufacturers decided that the rules had to be reduced to writing before the 1964 sessions. He detailed the various steps that were taken in arriving at the rules to govern the 1964 conference and read into the record the correspondence between the manufacturers and the Internationals on the subject. Grimes stressed that membership in the Manufacturers Association was binding and that the Association was the legal bargaining representative for its member mills. Then Grimes continued:

---

30. *Ibid.*
31. President-Secretary John Burke had retired shortly before this meeting because of ill health. His successor was Acting President-Secretary William H. Burnell, first vice-president, whose activities had been generally confined to eastern Canada.
32. "Minutes," p. 39.
33. *Ibid.*, p. 43.
34. *Ibid.*, pp. 43–44.

> It is our understanding that your two International Unions, United Papermakers and Paperworkers, and the International Brotherhood of Pulp, Sulphite and Paper Mill Workers, acting jointly as the party of the second part of our Uniform Labor Agreement, are here to meet with us as the sole, legal collective bargaining agency representing the defined employees who are employed by, and working in, one or another of the fifty Member Mills or plants heretofore named, and that you have the exclusive authority to negotiate the terms of a labor contract in their behalf.[35]

Vice-President Isaacson of the Pulp Workers and Vice-President Robertson of the UPP agreed that this was the case. William Perrin of the Employees Association and Pulp, Sulphite Local 68 then reviewed the meetings between the manufacturers and the Internationals and stated that what was done at the meetings was not necessarily wrong, but that the approach was wrong and had stirred people up. Then he read into the record the ground rules adopted by the prewage conference delegates from the local unions.

The manufacturers were rather puzzled by the presentation by Perrin. Apparently they did not know what to make of it. Sid Grimes commented:

> We are here to bargain. We have a signed agreement with the presidents of your two International Unions who are, as we understand, the duly constituted legal collective bargaining agency representing these mills. We have had no notice of the status of the group to whom you refer. I say "status"—as to their legal status, as to the fact that they are agents or representatives authorized by the legal bargaining agent, and the only thing I can say is that we are bound by the signed document that we have between the two bargaining agents, and those are the conditions under which we are here.[36]

Isaacson replied to Grimes stressing that he would adhere to the ground rules agreed upon by the Internationals, and Oscar Robertson made a similar statement. The negotiations then began in earnest.

The next day, April 23, the union delegation held its first caucus to discuss the sessions of the previous day. In opening the caucus, Ivor Isaacson stated:

> As one goes along in life he never misses a time when he has an assignment that is difficult to fulfill.
>
> Last Wednesday I spoke to you about the ground rules to be used. [I] made the statement, when the nominations were opened for officers, that the person elected chairman should abide by our rules. These rules were not followed.
>
> Some of you have been consulting with attorneys and so have we consulted ours. We tried to prevent you putting your ground rules on the floor, but a goodly number of you voted to accept them. I was shocked and surprised when Perrin read those rules into the record.

35. *Record of Negotiations,* April 22 to May 5, 1964, pp. 40–41.
36. *Ibid.,* p. 47.

> I have tried to be a friend to all of you. I have told you I am sympathetic to some of your statements but your approach is completely wrong.
>
> I have called our International Headquarters and have asked for guidance. I am not here to appease 150 delegates; I am here to make a settlement to establish a pattern for the West Coast.
>
> It has been the past practice that one of the delegates be elected chairman of the caucus, but some of you have disregarded past practice and I am going to do the same.
>
> I am now compelled to tell you, the three International Vice-Presidents are going to take over the entire bargaining and the caucus.[37]

After the statement of Isaacson the elected conference officers, led by Chairman Perrin, walked out of the caucus, followed by a large number of delegates. A roll call indicated that 70 delegates had left with Perrin and the other officers while 66 remained.[38] Rupture had occurred in the paper unions on the West Coast.

Appointments were made to fill the places on the bargaining board left vacant by the walkout, although any delegate who wished to return to the conference or resume his position on the bargaining board was permitted to do so.

While the International officers and staff and the delegates who remained at the conference continued their attempt to negotiate a new contract, the rebel group was busy attempting to halt the bargaining. The International Unions contended that the National Labor Relations Board, not the Courts, was the agency that had jurisdiction over the negotiations. That view was sustained on April 30, 1964,[39] and the conference continued.

The dissident group moved to assume an independent position while the conference was still in progress. The instrument was an organization entitled the ULA Committee, and was the vehicle for legal action during the month of May. It solicited the locals for funds in the fight against the Internationals. The Committee had been formed in October 1961, by members of the reform group on the West Coast and between February 12, 1962, and June 29, 1964, West Coast local unions donated some $160,167 to it.[40]

The rebel group, failing to halt the negotiations by injunction, took their case to the National Labor Relations Board, charging the Internationals with unfair labor practices. The regional director refused to issue a complaint.

---

37. "Minutes," p. 54.
38. *Ibid.*, p. 60.
39. *Ibid.*, p. 102.
40. Phillips et al. v. Perrin et al., Multnomah County, Oregon Cause No. 302–117, and Burnell et al. v. Perrin et al., Multnomah County, Oregon Cause No. 302–118, p. 2.

The negotiations continued, concluding on May 5, 1964. The manufacturers agreed to a contract which called for a 4 per cent general wage increase plus an additional 5 cent adjustment for women. The companies also agreed to contribute $2.50 per month for dependent health and welfare coverage and granted an additional paid holiday on July 3. The total package was estimated at 16¾ cents per hour. The unions agreed to a three-year contract, with wage and fringe benefit reopeners in each of the remaining years. The manufacturers had requested a three-year contract when negotiations first started and the unions had agreed. From the point of view of both parties, a contract running longer than the traditional one year was desirable. The unions would be given time to mend fences with the dissidents and the employers would be assured that a rival union would not enter the scene to complicate the situation.

## *The Formation of the Association of Western Pulp and Paper Workers*

On May 9, 1964, four days after the negotiations had concluded, the rebel group met in Olympia, Washington, and formed a rival organization in the pulp and paper industry on the West Coast—the Association of Western Pulp and Paper Workers. Thirty years of harmony was crumbling.

The immediate problem facing the rebel group was to secure rejection of the contract negotiated in Portland. If the contract had been accepted with its three-year term, the rebel group would have had no recourse for a time sufficiently long so as to allow the Internationals to attempt to restore harmony within their ranks. With a contract rejection, the rebels would be free to initiate proceedings looking towards disaffiliation from the Internationals and formation of an independent organization. The referendum vote to accept or reject the results of the Portland conference was concluded on May 22, and the contract was rejected by a decisive margin.[41] With the first hurdle cleared the rebels moved to establish their Association as the bargaining agent for the membership covered by the Uniform Labor Agreement. On June 1, the leadership of the Association, headed by William Perrin, petitioned the National Labor Relations Board to hold a representation election among the approximately 20,000 people covered by the ULA. In support of their petition the rebels presented the NLRB with about 9,500 signed authorization cards.[42]

After the rebels filed their petition for a representation election, the Internationals conducted a second referendum with a somewhat different

41. *Rebel* 1, no. 1 (May 1964). The unofficial count showed rejection by a margin of 9,128 to 5,214. The results announced by the International Unions favored rejection 8,858 to 6,729.
42. *Rebel* 1, no. 2 (June 3, 1964), p. 1.

question posed on the ballot. The choice given the employees was to accept the contract or reject the contract and strike. Thus, the question was posed in such a way that to vote against acceptance was tantamount to voting for a strike. The rebel Association advocated that the employees covered by the bargaining unit abstain from voting. In a light turnout, the results were 5,469 for acceptance with 543 favoring strike action.[43] The light turnout was a victory for the rebel Association. The Internationals requested that the employers effectuate the new agreement, but the employers association refused because of the pending representation petition.

By the beginning of July, the rebel Association claimed that 29 locals had disaffiliated from the Pulp Workers and the Papermakers. A "conservative" estimate placed the strength of the Association at 11,500.[44]

On July 13, the regional office of the NLRB held a hearing on the Association's request for an election and, in the early part of August, ordered that an election be held September 14 to 21, 1964. The International Unions appealed the ruling to the full Board in Washington, D.C. The appeal was denied on August 18, and the contest proceeded. Both the Internationals and the Association waged a vigorous campaign for the votes of the employees in the election to be held in September. The Internationals sent their top leadership to the West Coast in an attempt to build support. Both sides distributed handbills, held news conferences, issued press releases and published their own newspapers. The Internationals enlisted the support of the AFL-CIO and its regional bodies. However, in some instances local central bodies assisted the rebel organization in its fight against the Internationals. On July 13, George Meany, President of the AFL-CIO had telegraphed central bodies in the area covered by the ULA:

> I am advised that certain local central bodies in the area have permitted this secessionist group to obtain office space in central body facilities, have allowed members of this group representation at central body meetings, and have failed to render full support to the Pulp, Sulphite and Paper Mill Workers and the United Papermakers in their efforts to combat this destructive movement and to preserve the integrity of the bargaining unit as an AFL-CIO affiliate.
>
> You are herewith advised that any aid, assistance, cooperation or recognition of this secessionist group or its members is contrary to the best interests of the AFL-CIO and a violation of the duties and responsibilities of central bodies under the AFL-CIO Constitution and rules governing state and local central bodies.
>
> You are further advised that it is the duty and responsibility of every state and local central body to give all possible aid and assistance to

---

43. International Brotherhood of Pulp, Sulphite and Paper Mill Workers, "Memorandum Concerning Raid on West Coast AFL-CIO Paper Industry Unions" (no date), p. 1.

44. *Rebel* 1, no. 7 (July 8, 1964).

> the International Brotherhood of Pulp, Sulphite and Paper Mill Workers and the United Papermakers and Paperworkers in combating this rump organization of secessionists.[45]

The Pulp Workers also took a number of legal actions to help its cause. At an executive board meeting held in Glens Falls, New York, on July 8 and 9, the board placed 5 West Coast locals under trusteeships and instituted lawsuits for return of assets against 18 locals.[46]

During the months of June, July, and August, the manufacturers had maintained a position of neutrality in the dispute between the Internationals and the rebel Association. The provisions of the Uniform Labor Agreement were still in force as no party had terminated the contract. On September 9, the manufacturers abandoned their neutral stance and acted in a fashion which damaged the chances of the group they wished to support, the International Unions. On that date, each employee in all 49 member mills received a letter signed by the local mill manager suggesting that it was in the best interests of all concerned to maintain the current bargaining agent. The effect of the letter was to alienate many employees and substantiate the claims of the Western Association in regard to "collusion" between the Internationals and the employers. While the Internationals denied any knowledge of the letter or its motivation, the damage was done. Its effect could not be countered in the little time remaining before the balloting took place.

The election was held as scheduled on September 14 to 21 and the ballots were counted in Portland, Oregon, on September 23. The results were a victory for the Western Association. They won 10,653 to 8,130 for the Internationals. There were 91 ballots for no union, 134 challenged ballots, and 22 voided. Out of a total eligibility of 21,375 voters, 19,030 cast ballots. In a separate election for employees in Ketchikan, Alaska, not covered by the ULA, the Western Association won 254 to 34.[47]

## *The Western Association's First Contract*

Negotiations for an agreement between the AWPPW and the Manufacturers Association began in Portland on October 28, 1964. The companies evidently attempted to secure a greater freedom of action when they went into negotiations with the Western Association than they had been able to achieve with the Internationals. The manufacturers introduced a new set of proposals that looked towards eliminating the past record of negotiations and the introduction of a management rights clause. They also wished to introduce a weakened form of union

45. Contained in letter, William H. Burnell to all International officers, representatives and local unions of the International Brotherhood of Pulp, Sulphite and Paper Mill Workers, July 20, 1964.
46. *Ibid.*
47. *Rebel* 1, no. 18 (September 30, 1964), p. 1.

security. On November 3, the Federal Mediation and Conciliation Service entered the dispute. On November 12, pickets appeared at the plants of employer association members, though no unit-wide vote on a strike was conducted. Secretary of Labor Willard Wirtz called the negotiators to Washington and the AWPPW sent William Perrin, John Eyer, former representative of the Pulp Workers who had joined the new union, Elmer Lines, John Anglin, Fred Delaney, and Graham Mercer. Progress was made in the Washington talks and on November 24, the negotiators announced that they had reached agreement. The strike was terminated, and on December 8 the agreement was signed. It was ratified by a vote of 11,487 to 1,285.[48]

The contract negotiated by the AWPPW provided for a three-year pact with a reopener on March 15, 1966. In terms of cost to the employers, the AWPPW contract was about the same as that negotiated by the Internationals before the disaffiliation. The rebel contract provided for an additional one cent per hour on top of the increase already negotiated. In exchange for the penny, the Western Association gave up the $2.50 per month to be paid by the employers towards dependent health and welfare coverage. The provision on retroactivity of the contract terms limited back pay to employees on the payroll when the contract was signed. This had the effect of denying retroactive compensation to employees who were no longer in service, for example students who work in the mills as summer replacements. The Western Association took court action to secure retroactive pay, but in September 1966 a federal court denied the Association's claim and ordered it to pay the Manufacturers Association court costs in the case.

The Western Association was not able to secure as strong a union security clause as the International Unions had in the ULA they had negotiated. The Internationals had a clause that provided maintenance of membership for all employees hired prior to June 1, 1948, and a union shop for employees hired after that date. The clause negotiated by the Western Association provided that all employees hired on or after June 1, 1964, had to become members of the Western Association. It also provided that employees who belonged to the AWPPW on December 8, 1964, or who afterwards joined the union, had to maintain such membership. In addition, to cover employees not covered by the two preceding provisions, the contract provided 120 day period Local Mill Union Shop Option which stipulated that in cases where proof of 80 per cent or more membership was shown in that period from the date of the signing of the agreement a full union shop would become effective. This provision was satisfied in all member mills but one, the Crown-Zellerbach mill at Camas, Washington. The union security clause also provided that the

48. *Rebel* 1, no. 23 (December 9, 1964), p. 1.

Western Association could apply for an election in a member mill to determine the sentiment on a union shop at any time before December 31, 1965. An election was to be conducted by the Federal Mediation and Conciliation Service. If the vote showed 70 per cent of the total membership favoring a union shop, it would go into effect. Such an election was held at Camas, and the Western Association did not make the 70 per cent figure.

In the area of management rights, the contracts negotiated by the Internationals were silent. The document signed by the AWPPW introduced the subject in a clause that reserved to the signatory company all rights except as those rights were modified by the agreement. The clause stated that nothing in the agreement should be construed to impair the right of the signatory company to conduct its business except as expressly modified by the agreement. The rights of the union were specifically restricted to those rights conferred in the agreement.

## *The Legal Tangle*

During the month of May 1964, and in the last few weeks of April, the ULA Committee had solicited funds from local unions in order to finance the rebellion. The locals contributed about $162,000 to the Committee, the bulk of which was evidently used to finance the formation of the AWPPW. The International Unions took action to halt the dissipation of local union treasuries and sought the return of funds donated to the Committee. The argument of the Internationals stressed the forfeiture clause in their constitutions which provided that upon dissolution of a local its assets would revert to the International. In August 1964, the Internationals secured a temporary injunction freezing the funds donated to the ULA Committee which then amounted to about $17,000. The main case was the suit by the Internationals against William Perrin and the other officers of the Western Association, commonly known as the Portland Case. The argument of the Western Association stressed that in their opinion the forfeiture clauses in the International Constitutions should be considered inoperative since the Internationals had not fulfilled the fiduciary duty they owed the members of the local unions. The westerners maintained the Internationals came to the matter with unclean hands. The defense of the Internationals stressed that they had committed no violations of law and that many of the features that the westerners desired to see introduced into the Constitutions of the Internationals had been proposed in the various past conventions and had been defeated. The Portland case came to trial in late May 1965, before Judge Robert Jones, and lasted 32 days. A significant part of the AWPPW's case rested on the testimony of former International Representative Paul Hayes. After a lengthy deposition, the counsel for the Association requested that none of Hayes's testimony be introduced in the proceedings because of its "inconclusive" nature.

In his opinion, the judge had some words to say about Joseph Tonelli, who had been elected President of the Pulp Workers. Judge Jones commented:

> An examination of the character and activities of the Pulp Workers' President Joseph Tonelli leaves little doubt in an impartial mind why several thousand West Coast rank and file union members refuse to be affiliated with an international organization headed and controlled by such an individual.
>
> Mr. Joseph Tonelli manifests in his deposition testimony that is unworthy of belief.
>
> It is assumed that Mr. Joseph Tonelli did not have his head in the sand in respect to the importance of this litigation to his union. Yet, where was he? To the parties concerned, this litigation involves more than the mere dollars and cents prayed for by the plaintiffs. This suit has been of the utmost importance to many persons concerned with the labor movement. Implicit in this case is the question of whether or not local union members can successfully withdraw from a strong, centralized international power and run their own affairs through local control. Ultimately, the outcome of this litigation may have tremendous financial implications for the International Union. At stake may be the retention or loss of millions of dollars in contributions from West Coast union members. Yet, Mr. Tonelli, the present President of the Pulp Workers International Union, did not appear. As admitted by his lawyers the defendants main attack against the Pulp Workers centers around charges against Tonelli. Tonelli piously claimed in his deposition to be an "open book." Unfortnuately, he merely presented the preface, but ignored the major chapters.[49]

In respect to the main line of defense advanced by the AWPPW, that the Pulp Workers had breached their fiduciary duty towards their members, the judge had this to say:

> In spite of the adverse finding against plaintiffs in respect to Joseph Tonelli, on the whole there is not sufficient evidence of toleration of corruption within its ranks on behalf of the International Union or its duly elected officers to constitute such breach of a fiduciary duty as to justify avoiding the forfeiture clause.[50]

In reply to the claims of the defendant that the Pulp Workers had not provided the West Coast membership with sufficient service, the judge noted that the complete satisfaction of all who were represented was hardly to be expected. In his discussion of the assumption of power in the negotiating sessions that precipitated the breakaway by the western locals, the judge concluded that:

---

49. Phillips et al. v. Perrin et al., pp. 25–27.
50. *Ibid.*, p. 31.

> The plaintiffs, although violating historic concepts and practices for bargaining in the West Coast, were acting within their legal capacity as bargaining agents; and they had the legal right to meet with accredited representatives of the employer group. The agreement reached in reference to ground rules was also a legal agreement, within the authority of the bargaining agent. The action of the plaintiffs, taking over the bargaining to the exclusion of the West Coast delegates, was highhanded, arbitrary and in violation of the tradition and history of bargaining. Nevertheless, it was also an action that is legal and within the broad authority conferred upon the plaintiffs by law as bargaining agents. As the bargaining agents, the plaintiffs possessed not only wide responsibility but authority to meet that responsibility. . . .[51]

Thus, the judge ruled for the Internationals, requiring the Western Association to return approximately $160,000 to the two Internationals. The AWPPW posted a bond for the required amount and appealed the case. The result on appeal sustained the finding of the lower court.

The Internationals also took legal action against local unions that had left the fold. Foremost in these actions were suits against Oregon City, Oregon, Local 68, and Longview, Washington, Local 153. The suit against Local 68 involved $130,434 while that against Local 153 was for $104,997. The Longview case concluded in September 1966, with a victory for the local. The judge decided that the International had, indeed, committed a breach in its fiduciary relationship to the local which justified the local in taking steps to disaffiliate. The judge also decided that the International Union had a duty to conduct an independent investigation of Joseph Tonelli's affairs; and that the International was guilty of sufficient misconduct as to justify the nonapplication of the forfeiture clause in the Constitution. The judge declared the International came to court with unclean hands. In October 1966, a decision was rendered in the Oregon City case. The opinion of the judge in deciding for the local was similar to that read in the Longview case, with the addition of the point that in the opinion of the judge the International was guilty of failure to observe and carry out the mandate of the AFL-CIO Code of Ethics. Both cases were appealed.

In June 1968, a decision was rendered by the Washington State Supreme Court in the Local 153 case. The nine-judge court unanimously sustained the ruling of Judge W. R. Cole in finding for Local 153. The court noted that International Unions should not be permitted to use the forfeiture clauses in their constitutions as exploitive or punitive measures. Since the local would use its funds for their intended purpose, the Court declared it could not aid the International by enforcing the forfeiture provision of the Pulp Workers Constitution.

---

51. *Ibid*, p. 39.

These three cases represent the most significant legal actions to date. There have been at least thirty-three suits filed by one group or another as a result of the breakaway. The litigation may be divided into three categories: suits to recover assets filed by the Pulp Workers or the UPP; suits filed by the breakaway group to recover assets under the control of the Internationals or loyal officers; and libel and slander suits filed against the Pulp Workers or the UPP. At the very least, the breakaway represented a "full-employment act" for lawyers.

### *Recent Events*

In May 1967, the National Labor Relations Board conducted an election on the petition of the Pulp Workers and the UPP to regain representation rights on the West Coast. The AFL-CIO unions were decisively defeated by the Western Association, 14,448 to 2,609.[52]

When the AWPPW opened negotiations with the employers an impasse was reached, and on June 9, 1967, the delegates voted unanimously to submit the manufacturers' final offer to the membership referendum with a recommendation to reject.[53] When the ballots were counted the paperworkers had again demonstrated their independence by voting to accept the manufacturers' offer, 7,748 to 5,697.[54]

The definite rejection of the International's bid to regain bargaining rights plus the negotiation of a two-year contract which ran until March 15, 1969, seems to indicate that the Western Association has demonstrated its capacity for survival. It appears likely that the AWPPW will continue to exist for the foreseeable future.

---

52. International Brotherhood of Pulp, Sulphite and Paper Mill Workers, *Pulp and Paper Worker,* May 1967, p. 3.
53. *Rebel* 4, no. 18 (June 14, 1967), p. 1.
54. *Rebel* 4, no. 18 (June 28, 1967), p. 1.

# 8

# SCHISM AND ITS IMPLICATIONS

This study has attempted to set forth the circumstances surrounding the formation of an independent union in the pulp and paper industry on the West Coast of the United States. In view of society's expectation that unions conform to the democratic ideal, it is useful to appraise the events under study in light of their implications for union democracy.

The history of the Pulp Workers shows a fairly close correspondence with the theories of union development advanced by Richard Lester.[1] Lester divides his concept of union growth into three main categories: increasing centralization; changing status and outlook of union leaders; and a reduction of militancy within the union and the substitution of carefully prepared material in the collective bargaining process.

Under the heading of centralization of functions and control, the theory sets forth a number of factors that are instrumental in the shift of power from the local to the international level. Among them are the expansion of the union itself, the growth of the area of collective bargaining, the change in the character of the subjects bargained about, and the decline of rival unionism. To a greater or lesser extent, these forces have all been operative on the Pulp Workers. The union grew from an organization of a few locals concentrated in the northeastern part of the United States to an organization with locals in all sections of the continent. This growth reduced the influence of some of the early locals, such as Fort Edward Local 1, or Palmer Local 4. The increasing size of the union compels an expansion of the role of headquarters, if only in the area of administering the affairs of the organization. Staff is required and there is some opportunity to build up a political machine. In the Pulp Workers, staff was increased as the territory covered by the union expanded and the range of services demanded by the membership increased. Perhaps the greatest jump in the size of the union staff occurred in 1944 with the creation of the Department of Research and Education. It may be that the department was formed as a consequence of the threat of the rival CIO Paperworkers. If so, its creation was

1. Richard A. Lester, *As Unions Mature* (Princeton, N.J., 1958).

designed to increase the prestige of the International and perhaps the dependence of the locals.

In the area of collective bargaining the entire concept of the Uniform Labor Agreement on the West Coast operated to enhance the power of the International at the expense of the local. It is unrealistic to expect the locals, scattered as they are over three states and dealing with many firms, to have the perspective or point of view that the International is able to have. The area-wide nature of the ULA probably served to concentrate power in the hands of the top leadership of the Pulp Workers by reducing the importance and independence of the local and regional leadership. This tendency was reinforced by the actions of the International in 1942 when it successfully sought increased power and weakened the authority of the regional association, the Employees Association. From the point of view of the International, the wisdom of the area-wide negotiating format was demonstrated in the Hoquiam case decided by the NLRB during World War II. In that instance, a rival union, the Woodworkers, were denied an opportunity for a foothold in the coast-wide bargaining unit. Such a decision by the Board enabled the International to operate secure in the knowledge that it would be difficult, if not impossible, for a rival union to challenge its position. The Board's decision also reduced the options available to the locals since they could not threaten to leave the bargaining unit and join a rival union. The merger of the CIO Paperworkers with the AFL Papermakers in 1957 further restricted the area of choice available to the locals. The feeling of security that the development of the ULA may have produced in the officialdom of the Internationals may have operated to reduce the vigor with which collective bargaining demands were prosecuted. The emphasis of the central administration was on the preservation of the ULA, and on maintaining harmonious relations with the companies. Secure, the Internationals could afford to act more like labor statesmen and less like the militant negotiators their membership desired.

Lester makes the point that the militancy of the trade union movement has declined in the past twenty or thirty years. Unions are more disciplined and businesslike. They are less militant, says Lester, because of the centralization referred to earlier, and because of the impact of several other developments. Lester mentions greater employer acceptance, the increase in the workers' living standards, and the middle-of-the-road attitude of American society in recent years. In the area of union militance, it would be fair to say that the unions in the pulp and paper industry, the Pulp Workers and the UPP, do not have a reputation for being especially vigorous. They have accommodated themselves to the employers in the industry. In large part, this is probably a reaction to the strike against the International Paper Company in the 1920s. In that conflict the unions took on the company along a wide front and were decisively defeated. Since those years the unions have been reluctant to strike a company with nationwide or continent-wide operations. This does not mean the unions have been unwilling to strike. They have struck, but only when they felt forced to by the attitude of the firm with whom they were dealing. The Pulp Workers and the old AFL Papermakers were willing to deal with the firms in the industry on the basis of mutual accommodation. Thus, the union granted wage concessions during the depression of the 1930s and the employers responded, in many cases, by not opposing the entry of the unions into their plants. The same procedure was evidently operative during the 1950s when some firms actively aided the organizing campaigns of the Pulp Workers on the West Coast. For many firms, the Pulp Workers and the Papermakers are a known quantity with whom relations have been maintained for many years. A different union would upset the pattern that has developed. The Pulp Workers do not press too hard for wage increases and, in most cases, do not have to do so. The high productivity of the industry, coupled with its generally good profit position, has enabled it to grant wage increases that evidently satisfy the membership of the unions. With most of the industry organized and with no significant competition from rival unionism for most of the period under study here, it is understandable that the unions may not have felt any pressure to exploit their economic strength to the utmost. It was easier to accept the gains that came without strike.

Apparently the Executive Board of the Pulp Workers was not aware of the seriousness of the situation or the degree of dissatisfaction that existed on the West Coast. During the 1962 Convention the West Coast area was forced to accept a vice-president who was clearly not the choice of the locals in the region, yet delegates from the other sections of the continent and the executive board voted for him. The same situation occurred in British Columbia where all the locals in the area supported the challenger to the incumbent vice-president and saw their desires thwarted by the remainder of the continent and the executive board. The only supporter of the reform group on the executive board faced a challenge for re-election in 1962, and the challenger received the support of a majority of the executive board. The evidence seems to indicate that the board was attempting, successfully, to limit dissent within its ranks. Despite all the claims that the locals could have the vice-president they desired without instituting a system of area-wide voting, the facts show clearly that the claims were not justified. Two choices of different regions were defeated, and an incumbent vice-president who was clearly favored by the people he served almost went down to defeat with his opponent supported by the executive board.[2] In insulating itself from the rank and file member by stifling dissent within its ranks, the executive board was preparing the ground for the revolt of 1964. The board was composed of men who thought alike on major issues and apparently underestimated the feelings of the membership on the West Coast on the issues of internal democracy and regional autonomy. The evidence seems to indicate that the revolt was not inevitable. Judicious handling of the locals on the Coast could have prevented the rupture. However, the members of the executive board were not prepared to cede any authority to the locals, or even let the locals exercise the role that was historically theirs. Apparently the International hierarchy was more interested in maintaining good working relationships with the companies, and in preserving the bargaining structure, the ULA, than in responding to the wishes of their membership. In order to achieve these objectives, the administration was willing to comply with the manufacturers' desires for a reduction of the role of the locals. In addition, they negotiated with the companies in regard to the ground rules in February 1964 without the knowledge of the locals directly affected. Upon learning of the International's virtual acceptance of the employers' ground rules the locals were understandably disturbed. Even during the negotiations of 1964, the situation could have been remedied. If the International officers present had been less highhanded in their conduct and had backed off from their rigid stand on the question of who was to have the final authority in negotiations, an accommodation might still have emerged. By bluntly stating that the International hierarchy was going to take over the negotiations Vice-President Isaacson forced the revolt. Isaacson's action left William Perrin, the reform leader and leader of the locals, with two choices. He could accept the position of the Internationals, or he could fight to preserve the historic role of the West Coast locals. Perrrin chose to fight. At that point the rebellion became a fact. The action of the Internationals in attempting to usurp authority that had traditionally resided in the locals practically forced Perrin to make his stand.

The question arises concerning whether or not the revolt was motivated by the efforts of the West Coast leaders to secure power within the International or whether it was a sincere attempt to assert what the westerners regarded as the principles of democratic trade unionism. If the situation under study here was purely an attempt of disgruntled local leaders to gain power, it seems likely that the revolt would not have taken place in 1964. Rather, it would have commenced in 1962 after the defeat of the RFMDA in the convention. That was the logical

2. The vice-president in question, Godfrey Ruddick, merely delayed his ouster for 3 years. The 1965 convention defeated him in his bid for reelection.

time for a rebellion to begin. The reformers had been defeated on every issue in Detroit, but they still had a viable organization that could have worked for a decertification election in 1963 when the contract expired. This did not occur. The verdict of the convention was accepted by the West Coast locals, with the exception of some Canadian locals in British Columbia who left the International. The United States locals, however, apparently made a sincere attempt to work with their vice-presidents. The RFMDA withered in the period from September 1962 to April 1964. The reform group was better organized and financed in 1962 than it was in 1964. There was apparently no advance preparation for the disaffiliation moves of April 1964. By attempting to run roughshod over the locals, the Internationals precipitated the events they were seeking to avert.

The West Coast leadership stood for a type of trade unionism that was somewhat different from the traditional Pulp Worker orientation. The Pulp Workers have historically vested considerable power in the president and the executive board. The constitution of the union provides that the president shall appoint international representatives and other employees. It contains a lengthy section on imposition of trusteeships and has a detailed clause relating to forfeiture of local union assets. The forfeiture clause provides that, in the event a local union is suspended for violation of the constitution, all books, monies, and property of the local revert to the International.[3]

In keeping with their concept of a democratic trade union, the Western Association drafted a constitution considerably different from that of the Pulp Workers. Operation of the union is quite decentralized. The AWPPW Constitution established five areas and provides that trustees will be elected from each. Representatives are also elected under the AWPPW Constitution. In addition the constitution provides that area councils should be established which have the right to submit resolutions to the AWPPW convention. The constitution provides a recall procedure for executive board officers and area representatives and trustees. In the event a local loses its charter, the AWPPW Constitution stipulates that the assets of the local remain the property of the local. The AWPPW Constitution carries out the concept held by the westerners that the trade union should be decentralized and democratized to the greatest extent possible. The constitution is the best demonstration of the sincerity of the independent group. They have put into practice what they preached to the International. The reform group was evidently sincere in its desire for a change in the method of operating the International. Their constitution reveals a profoundly different conception of how a

3. International Brotherhood of Pulp, Sulphite and Paper Mill Workers, *Constitution and By-Laws.*

trade union should be governed. It seems to demonstrate that the reformers did not raise the issues of regional election and establishment of a review board as mere political ploys, but rather as sincere attempts to change the structure of their International. When they felt compelled to leave their union they put into practice the features they felt insured a democratic trade union.

### *The Role of the Manufacturers*

In the establishment of the reform caucus and the eventual revolt, the manufacturers played a significant role. In their attempts to continue relations with the Internationals, they almost insured that the ferment on the Coast would become revolt.

In the 1959 negotiations for renewal of the ULA, the employers intransigence on the pension issue gave the Pulp Workers Research and Education Director George Brooks an opportunity to advocate a strong position in opposition to the position of the International administration. By giving Brooks the opportunity to adopt a different position, the employers to some extent enhanced his status in the eyes of the locals. By precipitating action on this issue, the companies exposed the position of the Internationals and showed their reluctance to press a claim ultimately upheld as valid. Consequently, they furnished the reformers within the unions ammunition to make charges that the Internationals were more interested in the wishes of the employers than in the desires of their members. They also placed George Brooks in the position of appearing to be the only party interested in the welfare of the workers. He appeared to be fighting both the companies and the Internationals in his statement that a charge would be filed on the issues of pensions. When Brooks was fired in 1961, the reform elements, which until then had not had an issue to rally behind, were given a golden opportunity.

The manufacturers miscalculated again in 1963 and 1964 by insisting on the adoption of written ground rules to govern the conduct of negotiations. In the thirty years of harmonious relations between the manufacturers and the Internationals, the rules for conduct of the negotiating sessions had never been reduced to writing. By insisting on codification of the rules, perhaps with a view to helping the Internationals strengthen their position, the employers provided the spark to ignite the rebellion. The manufacturers were aware of the factionalism within the unions on the West Coast, but they evidently felt that the International officers could control the dissident group. This assessment of the situation proved incorrect.

In the final days of the election campaign, the employers blundered once again by contacting their employees and advising them to vote for the Internationals. There is no way of knowing how many votes were changed by this employer tactic, but if the contest was close, the letter may have provided the edge for the Western Association by lending substance to their charges that the Internationals and the companies were cooperating too closely.

The evidence seems to indicate that the companies wanted to continue their relationship with the International unions. They had good reason for wanting to do so. There had not been a strike in the bargaining unit for the thirty years it had been in existence. The Internationals did not press too hard in collective bargaining. The companies on the Coast enjoy a good profit position, and they could afford to give good settlements without the unions having to be especially vigorous.

The employers, along with the unions, must share responsibility for perpetuating a system of collective bargaining that had perhaps outworn its usefulness. As the Uniform Labor Agreement existed in 1964, it consisted of many thousands of pages of interpretations, the contract itself, and the transcript of negotiations. The record that had developed over the years involved such masses of material that it is doubtful that more than a handful of persons, if that many, completely understood the contents of the contract and its ancillary materials. Perhaps this situation was not uncomfortable to the parties. It was, however, unsatisfactory to the workers in the locals who had to live under the agreement. Control of administration of the agreement had passed from the locals to the International over the years and created a situation the locals did not care to continue. Their drive for more regional autonomy and authority clashed with the desire of the Internationals to continue the existing situation which centralized power in their hands. The Internationals were supported in their attitude by the companies. The last thing the manufacturers wanted to see was the return to the locals of some of the power that had come to reside at the International level. Dealing with the International staff was much more satisfactory for the employers. Somehow certain local demands that reflected changing local conditions always got lost during the negotiating sessions. Examples are the attempts the locals initiated for changes in the pension structure, addition of a converting supplement and a revised arbitration clause designed to bring the grievance procedure back to the local level. The Internationals supported the employers in the attempt to keep the locals from playing a meaningful role in the negotiation and administration of the ULA. During the mid-1950s, the vice-president on the West Coast had warned President Burke that the activities of the Employees Association would break up the ULA. The pressure of the locals on the pension issue in 1959 was also condemned for its alleged tendency towards destroying the ULA. The Uniform Labor Agreement became a sacred cow for both the manufacturers and the Internationals, and the interests of the locals were sacrificed to perpetuate it.

## *The Role of the Internationals*

That the Pulp Workers and the UPP hierarchy had a peace at any price approach to their relationship with the employers seems evident from the record presented in this study. The continual emphasis on maintaining the bargaining unit, along with the feeling of security produced by the NLRB decision of 1943 in the Hoquiam case, evidently made the Internationals insensitive to the desires of the rank and file on the Coast. Three times in the decade preceeding the revolt, 1953, 1957, and 1963, the local membership had rejected the terms reached in the initial round of negotiations. In all three cases, when union negotiators were instructed to go back for more, the final results varied little from the employers' original "last offer," and the terms were finally accepted by the unions. In spite of the criticism from the locals, in spite of the rejection of the contracts by the membership, a strike vote was never taken. Such a vote, a standard negotiating technique, might have buttressed the union position in its dealings with management. The fact that such a tactic was not employed, in fact seemingly not seriously considered, reveals the timorous attitude of the unions. This attitude was apparently restricted to the hierarchy. The rank and file was more militant. It was rank and file pressure that forced the Internationals, against their will, to prosecute the unfair labor practice charge on the pension issue. The removal of George Brooks from his position with the union, coupled with the victory of the unions before the Board and the Court of Appeals confirmed for the rank and file their feelings that a militant stance could produce gains for them. It also showed that the Executive Board of the Pulp Workers was determined to maintain its position, even at the expense of the membership. The rising chorus of criticism of the Internationals from the mid-1950s is also indicative of the position of the Internationals. Rather than taking advantage of the attitude of the Coast membership in an effort to get concessions from the employers, the entire thrust of International activity seems to have operated to keep the militants in check in order to preserve harmony with the employers.

The attitude of the Internationals is reflected in the settlements reached with the Employers Association. The West Coast locals lost ground from 1944 to 1962 to the other large regional settlement—that reached with the Southern Kraft Division of International Paper Company. In 1944 the male base rate under the ULA was $0.90; Southern

Kraft's was $0.635, a difference of $0.265 or 29 per cent. By 1952 the rate on the West Coast was $1.72 while that in the South was $1.32. The absolute differential had widened in favor of the Coast workers who were $0.40 or 24 per cent ahead of their Southern counterparts. But by 1962 the West Coast rate had advanced only to $2.295 while Southern Kraft's had advanced to $2.01. The absolute differential had fallen to $0.285 and the percentage to only 13. Clearly the union was more successful in the South. With similar economic conditions in both sectors, economic factors do not seem to provide the explanation.[4] This study indicates that the lack of militancy on the part of International offices on the Coast was a prime factor in the revolt.

As the International was apparently attempting to reach an accommodation with the companies in the face of membership resistance, its efforts to increase its power at the expense of the locals further exacerbated the situation. The entire thrust of International activity since the formation of the bargaining unit in 1934 had operated to reduce the role of the locals and the Employees Association. There was a corresponding increase in the power of the International. Originally the Employees Association had led negotiations with the employers. In 1937, the International officers were given an increased role in dealing with the companies. In 1942, the Internationals, with the grudging acquiescence of the Employees Association, was give the power to call the prewage conference and the authority to conduct the negotiations with the Employers Association. The Board decision in 1943 in regard to Pulp, Sulphite Local 169 at Hoquiam had the effect of making the position of the Internationals almost impregnable by making the entire membership of the Employers Association the relevant bargaining unit and preventing the employees at one mill from leaving. At the 1956 Convention of the Pulp Workers, the passage of a constitutional revision further reduced the role of the locals by deleting the clause which gave the wage conference power to formulate policy towards the manufacturers.

The continuing additions to the power of the International which left the locals a smaller role to play, coupled with the lack of an aggressive, militant position towards the employers, produced the revolt of 1964. The immediate causes were the agreement with the manufacturers on ground rules for conduct of the negotiations and the dictatorial attitude adopted by Isaacson and the other officers at the conference. But these were merely the culmination of years of frustration for the locals who had chafed under the reduction of their power and the poor contracts that had been negotiated. Witnessing the last vestiges of their authority

4. Harold M. Levinson, *Determining Forces in Collective Wage Bargaining* (New York, 1966), p. 131.

vanishing under the agreements reached by their International officers and the companies, the rank and file took the road towards independence.

## *The Current Situation*

With the experience of three contracts behind them, it seems likely that the companies are not at all upset now with the installation of the AWPPW as the bargaining agent. In some respects, the independent has less bargaining power than the Internationals. Although there was a strike during the negotiations for the first agreement in 1964, the basic position of the Western Association is weak. They do not have locals elsewhere on the continent and, consequently, have somewhat less power to hurt the firms with whom they deal. Most of the companies in the Manufacturers Association have operations spread geographically throughout the United States and Canada and thus are not as vulnerable to economic pressure as firms with a narrow geographic base. While the AWPPW is probably more militant and more responsive to local desires than the Internationals were, its room for maneuver is limited by the geographical position in which it finds itself. For the present, the great advantage of the Association is its responsiveness to local needs. Its constitutional structure ensures that the officers in the field and at headquarters are close to the people they represent. With their heritage of attention to local affairs, it seems unlikely that in the near term future that the union administration will become remote from the men in the mills. The main strength of the Association in collective bargaining may be in the "arm's length position" of the union in respect to the companies. During the time the manufacturers maintained relations with the Internationals, there developed between the two groups a certain feeling of *camaraderie*. This spirit was reflected in the fact that the unions and the manufacturers maintained an office in the same building in Portland, Oregon, and telephone calls were handled on the same switchboard. The officials on the manufacturers' side and on the union's side, in a certain sense, developed together. They had been associated with each other for many years, and perhaps there developed too much of a desire to see the other side's point of view. This was especially true of the unions.

The officers of the Western Association do not share the heritage of thirty years of relations and consequently may be less disposed to accept the viewpoint of the manufacturers.

In assessing the development of the Western Association it must be mentioned that it was not due, to any great extent, to the venality or corruption of International officials. President Burke was not a trade union hoodlum, nor were the officials on the West Coast. The Pulp Workers did not resort to gang warfare or coercion to enforce their point of view. The Internationals have never known corruption to any great extent. They were cited twice in the McClellan Committee hearings that led to the passage of the Landrum-Griffin Act, and then only in a peripheral fashion. Neither the Pulp Workers nor the Papermakers have ever stood convicted of the mournful litany of crimes that some other unions and their officialdom have been accused of committing. The allegations made against the current Pulp Workers President, Joseph Tonelli, were in large part designed to gain support for the reform program, and later the formation of the independent union. The alleged improprieties of International officials were only a device to gain support for the basic objective, more regional autonomy within the Internationals, and failing that, to gain independence. The membership was not denied its rights under the union constitution, funds were not embezzled or misappropriated, nor was there suppression of the members' right to speak against their officers.

## *Implications for Collective Bargaining*

The revolt of some 20,000 workers in the pulp and paper industry on the West Coast of the United States is part of the current ferment in industrial relations. The years since the close of World War II have seen our industrial relations systems in the United States moving in two directions. One has been toward more centralization of decision making power and the expansion of bargaining units. Opposed to this trend, there has developed more recently a reaction which seeks increased decentralization and a greater degree of local autonomy in negotiating and administering the labor agreement. These two forces clashed in the situation under study here and, at least for the short run, decentralization emerged the victor. The trend toward increasing pressure from the local level, for the resolution of local problems, has not been confined to the Pulp Workers by any means. Recent years have witnessed an increase in the number of instances where the rank and file have failed to ratify a contract negotiated at the national or regional level. There have been wildcat strikes in the auto and bituminous coal industries over the failure of the national negotiations to resolve local problems. The leaders of several major unions, the Steelworkers, the Electrical Workers, and the Rubber Workers, have been defeated recently in re-election attempts, and in all instances the charge was heard that the leadership had lost touch with the rank and file. Clearly, the West Coast workers in the pulp and paper industry were not acting in isolation, but as part of the current ferment in the trade union movement which seems to have developed in large part as a reaction to the great, and growing, power of the national union over its affiliated locals. The central problem in this area is the need to maintain that degree of centralization which will ensure that the union can confront the employer with enough power to make bargaining a reality and at the same time heed the needs and desires of local unions and special interest groups for effective representation of their peculiar claims. This study indicates that the demands of special interest groups or local unions are likely to be rejected when the administration of the International Union believes that the membership has little option except to accept the dictates of the hierarchy. A union schism of this size had never taken place before, and it is reasonable to believe that the International officers felt their membership was locked in and had little or no chance of winning a representation election. It would be easy to advocate increased decentralization of decision-making power on the basis that it would promote internal democracy. At the same time, we could also argue for an expansion of the powers of the International over the local on the grounds that it promotes stability in industrial relations. But these would be oversimplified solutions to a complex problem. For the union, decentralization of its authority might mean it would lack the strength to force concessions from its opponent across the bargaining table. The advantages gained by the local and the rank and file might evaporate under the massed power of the employer. What seems to be needed is action by the unions to give special interest groups a greater voice in internal union affairs. The Pulp Workers suppressed the pressures for change within the union, at great cost to themselves. Other unions have shown a more realistic appraisal of the situation. For example, some unions have modified their internal structure to give special interest groups better representation within the councils of the organization. The UAW has a department for its skilled tradesmen, its women members, and its technical, office, and professional membership. The Steelworkers, too, have moved to give their white collar members more representation within the councils of the union.

Another fruitful approach to the problem might involve leaving certain issues unsettled at the national or regional level and letting the local unions dispose of them at the plant level. This approach is followed in the automobile industry. In the steel industry, the 1965 round of negotiations allowed for local union participation on local issues. If the Pulp Workers had been less ready to deal with management and more willing to heed the desires of their local unions on the West Coast, a similar system could have been worked out in that situation.

This study seems to indicate that in this instance, when an International union was not responsive to the desires of its membership, the legal and bargaining structure of industrial relations was flexible enough to permit the will of the membership to be expressed. It is of great importance that this flexibility be retained. Public and union policy must preserve the right of the rank and file to dissent against its established leadership. While this tolerance of dissent may produce conflict and strife, it is more compatible with a democratic system than promotion of central power which stifles local desires. The West Coast rebellion should serve as a warning to the trade union movement that centralization of authority, without the consent of the membership, may not be tolerated. The failure of the International unions, with the support of the AFL-CIO, to bring the revolt to an end indicates that under favorable circumstances a determined group can indeed forge its own destiny.

# BIBLIOGRAPHY


## Books

Bernstein, Irving. *A History of the American Worker, 1920–1933: The Lean Years.* Boston, 1960.

Kerr, Clark, and Randall, Roger. *Causes of Industrial Peace Under Collective Bargaining, Crown Zellerbach Corporation and the Pacific Coast Pulp and Paper Industry.* Washington, D.C., 1948.

Larrowe, Charles P. *Shape-up and Hiring Hall.* Berkeley, Calif., 1965.

Lester, Richard H. *As Unions Mature.* Princeton, N.J., 1958.

Levinson, Harold M. *Determining Forces in Collective Wage Bargaining.* New York, 1966.

MacDonald, Robert M. "Pulp and Paper." In *The Evolution of Wage Structure,* edited by Lloyd G. Reynolds and Phillip Taft. New Haven, Conn., 1956.

Selvin, David F. *Sky Full of Storm: A Brief History of California Labor.* Berkeley, Calif., 1966.

Sherman, John. *Twenty Years of Collective Bargaining and Twenty Years of Peace.* Glens Falls, N.Y., 1954.

Troy, Leo. *Trade Union Membership, 1897–1962.* New York, 1965.

Widick, B. J. *Labor Today.* Boston, 1964.

Wolman, Leo. *Ebb and Flow in Trade Unionism.* New York, 1936.

## Periodicals

Brooks, George. "Reflections on the Changing Character of American Labor Unions." In *Proceedings of the Ninth Annual Meeting of the Industrial Relations Research Association.* Madison, Wis., 1957.

Brooks, George W., and Allen, Russell. "Union Training Programs of the AFL Paper Unions." *Monthly Labor Review,* April 1952.

Gross, James A. "The Making and Shaping of Unionism in the Pulp and Paper Industry." *Labor History,* Spring 1964.

Maclaurin, W. Rupert. "Wages and Profits in the Paper Industry." *Quarterly Journal of Economics,* February 1944.

## U. S. Government Publications

U.S. Bureau of Labor Statistics. *Directory of National and International Unions in the United States, 1965,* Bulletin No. 1493. Washington, D.C., 1966.

U.S., Congress, Senate, Select Committee on Improper Activities in the Labor or Management Field. *Hearings, Part 10.* 85th Cong., 2nd sess., 1958.

## Union Publications

International Brotherhood of Paper Makers. *Proceedings of the Fourteenth Convention,* March 6–10, 1939. Albany, N.Y.

———. *Proceedings of the Special Convention*, March 4–5, 1957. Albany, N.Y.

International Brotherhood of Pulp, Sulphite and Paper Mill Workers. *Constitution and By-Laws.* 1965.

———. *Executive Board Report*, January 1954.

———. *Executive Board Report*, January 1955.

———. *Letter from William H. Burnell to all International Officers, Representatives and Local Unions*, July 20, 1964.

———. *Memorandum Concerning Raid on West Coast AFL-CIO Paper Industry Unions*. No date.

———. *Minutes of the Meeting of the Executive Board*, April 1–3, 1960.

———. *Minutes of the Meeting of the Executive Board*, July 6–8, 1960.

———. *Pacific Coast Pulp and Paper Mill Employees Association, Minutes*, January 1958.

———. *Proceedings of the Tenth Convention*, October 3–5, 1922. Fort Edward, N.Y.

———. *Proceedings of the Eleventh Convention*, October 7–9, 1924. Fort Edward, N.Y.

———. *Proceedings of the Twelfth Convention*, October 5–8, 1926. Fort Edward, N.Y.

———. *Proceedings of the Thirteenth Convention*, March 5–7, 1929. Fort Edward, N.Y.

———. *Proceedings of the Seventeenth Convention*, March 16–19, 1937. Fort Edward, N.Y.

———. *Proceedings of the Eighteenth Convention*, March 14–17, 1939. Fort Edward, N.Y.

———. *Proceedings of the Nineteenth Convention*, September 8–12, 1941. Fort Edward, N.Y.

———. *Proceedings of the Twentieth Convention*, October 9–13, 1944. Fort Edward, N.Y.

———. *Proceedings of the Twenty-second Convention*, August 10–19, 1950. Fort Edward, N.Y.

———. *Proceedings of the Twenty-fourth Convention*, September 24–29, 1956. Fort Edward, N.Y.

———. *Proceedings of the Twenty-fifth Convention*, August 31–September 4, 1959. Fort Edward, N.Y.

———. *Proceedings of the Twenty-sixth Convention*, September 10–16, 1962. Fort Edward, N.Y.

———. *Pulp and Paper Worker*, May 1967.

———. Rank and File Movement for Democratic Action, *Handbook for Constitutional Reform*, August 1962.

International Brotherhood of Pulp, Sulphite and Paper Mill Workers, and United Papermakers and Paperworkers. *Minutes of the Pre-Wage Conference*, April 1964. Portland, Ore.

International Brotherhood of Pulp, Sulphite and Paper Mill Workers, United Papermakers and Paperworkers, and Pacific Coast Association of Pulp and Paper Manufacturers. *Record of Negotiations*, May 15–29, 1959. Portland, Ore.

———. *Record of Negotiations*, May 3–24, 1963. Portland, Ore.

———. *Record of Negotiations*, July 9–20, 1963. Portland, Ore.

———. *Record of Negotiations*, April 22–May 5, 1964. Portland, Ore.

United Paperworkers of America. *Proceedings of the Second Constitutional Convention*, October 2–5, 1950.

———. *Officers Report to the Fourth Constitutional Convention*, April 25–29, 1955.

———. *Proceedings of the Special Convention*, March 4–5, 1957.

## Union Journals

International Brotherhood of Paper Makers. *Paper Makers Journal.*

International Brotherhood of Pulp, Sulphite and Paper Mill Workers. *Pulp, Sulphite and Paper Mill Workers Journal.*

## Union Newspapers

International Brotherhood of Pulp, Sulphite and Paper Mill Workers. Rank and File Movement for Democratic Action. *The Amplifier.*

Association of Western Pulp and Paper Workers. *The Rebel.*

## Unpublished Material

Madison. Wisconsin State Historical Society. Papers (1906–1957) of the International Brotherhood of Pulp, Sulphite and Paper Mill Workers.

Madison. Wisconsin State Historical Society. Papers of the Rank and File Movement for Democratic Action (1960–1963), International Brotherhood of Pulp, Sulphite and Paper Mill Workers.

Brotslaw, Irving. "Trade Unionism in the Pulp and Paper Industry." Ph.D. Dissertation, University of Wisconsin, 1964.

Kattan, Leonard. "The Private Settlement of Industrial Disputes." Unpublished Paper. Graduate School of Business Administration, University of California at Los Angeles, January 1966.

## Letters in Possession of the Author

Paul Phillips to John Burke, July 21, 1959.

John Burke to Paul Phillips, July 17, 1959.

Fred Delaney to John Burke, August 18, 1959.

Clarence Dukes to John Burke, August 21, 1959.

Ralph E. Davison to John Burke, September 29, 1959.

# INDEX

Abel, I. W., xiii
Abitibi Pulp and Paper Co., 8 n.55, 10
Advance Paper Company, 12
AFL-CIO, xi, xiii, 103, 138, 144
Albany Corrugated Container Corporation, 110
Albert, Maurice, 78
Amalgamation of unions, 44, 46. *See also* Merger
Ameden, Richard H., 76, 85, 93
American Federation of Labor, 5, 22–23
American Paper and Pulp Association, 15, 33
*Amplifier*, 90, 94, 95–96, 99–100, 116
Andawagan Paper Products Company, 17
Anglin, John, 140
Anglo-Canadian Paper Company, 10
Anonymous memorandum, 117–21, 126
Antioch Local 249, 45, 48, 50, 85 n.45, 89
Appeals board, 108–109, 152
Arbitration procedure, 121, 151
Archer, Dick, 132–33
Arkadelphia Local 947, 109
Ashe, David, 78
Association of Catholic Trade Unionists, 53, 58, 80
Association of Western Pulp and Paper Workers (AWPPW), 137–44; current status of, 154; decisive victory of, 144, 151, and democratic unionism, 149; first contract of, 139–41; founding of, 137–39; membership of, 1

Ballantyne, James, 85 n.45
Barbaccia, Anthony, 53, 58–59, 71, 81, 111–12
Bargaining board, 129–31
Barnes, Frank C., 53, 72, 74; resignation of, 68, 76–77
Bartlett, Harold, 85 n.45
Bellingham Local 194, 66, 101, 123
Bellingham Local 309, 36
Berlin Local 23, 4
Bernstein, Irving, 11
Bogalusa Paper Co., 16
Braaten, Orville, 45 n.9, 67–68, 81, 85, 86 n.48; alleged communist tendency of, 94; and Detroit convention, 102; and Pulp and Paper Workers of Canada, 116; and RFMDA, 93
Bradford, Raymond, 54, 84 n.41, 89
Bradshaw, Joseph, 84 n.41
Brooks, George C., 14
Brooks, George W., 43; charges against, 82; later views of, 118 n.48; and pension plans, 62–65, 150; and research and education department, 45, 47–48; resignation of, 80–85, 150, 152; and RFMDA, 86–92, 102, 114; and Uniform Labor Agreement, 62–65, 150
Brown, Al, 58, 122
Brown Company, 23
Brunk, Reeves H., 113
Brunswick Local 400, 96, 108
Bryce, R., 113
Bureau of Labor-Management Reports, 110–112
Burke, John P., 8 n.58, 9; and anonymous memorandum, 117; and Barbaccia, 59, 111; and 1959 convention, 68; and Crown-Zellerbach, 54–55; and depression, 11, 13–14; and Detroit convention, 102, 104–105, 107, 115; and Employees Association, 36; and International Paper Co., 7; and McNiff, 93; and minority groups, 59; and organization of Canada, 10; and organization of Lake States, 17–18; and organization of Northwest, 25–33; and Program for Militant Unionism, 75–76; and proportional representation, 53; questionnaire to, 77–78; and regional elections, 52; retirement of, 134; and rival unions, 23; and Ruddick's reelection, 113; and St. Regis Paper Company, 6; and Sherman, 35–37, 44; and training programs, 44–46; and *Truth*, 90–91; and unfair labor practice charge, 65; and Uniform Labor Agreement, 49–50, 53; and union constitution, 37
Burkey, George, 28–30
Burnell, William, 13, 15, 44, 51; as acting president, 134; opposition to, 68
Burns, Matthew, 11, 14, 32

Camas Local 100, 106
Canadian International Paper Company, 18
Capital Local 230, 27
Carey, J. T., 4–5, 7, 8 n.58
Carlisle, Floyd D., 8 n.58
Carlisle, Fred, 6
Carlisle group, 6, 8–9
Centralization, 41, 118, 145–46, 155–57
Chatham, R. H., 85–86, 95, 115
CIO, 19–22
CIO Paperworkers, 3 n.14, 5
Cliff Paper Co., 8 n.55
Clute, William, Jr., 107
Code hearings, 32
Cole, Judge W. R., 143
Collective bargaining, xiii–xv, 145–46, 151; and constitutional change, 53; and Employees Association, 44; implications of, 155–57; International position on, 130; in the Northwest, 41; and pensions, 61–63; in reform plan, 75; and West Coast split, 117–18, 120
Columbia River Paper Mills Co., 34 n.40
Combined Locks Paper Company, 17
Communist subversion, xii; and Barbaccia, 59; and longshore strike, 31–32; and RFMDA, 94–95; and Scarselleta, 94–95
Compulsory retirement of board members, 67, 75
Conference of 1959, 61–66, 82
Connolly, Patrick, 68, 71; and Brooks resignation, 82; dismissal of, 73–74, 76–77; possible reinstatement of, 96 n. 99, 114; and *Truth*, 90–91. *See also* RFMDA
Consolidated Water Power and Paper Company, 17
Constitutional change, 53; and convention of 1956, 153; and Portland case, 141; and Program for Militant Democratic Unionism, 74–75; and RFMDA Handbook, 100 102
Constitution Committee, 67–68
Continental Bag Company, 6
Contract referendum, 67, 137–38
Convention at Detroit (1962), 99–120, 149
Convention at Montreal (1959), 67–69
Convention, founding, 3
Convention of 1906, 4
Convention of 1956, 47, 53
Convention of 1965, 51 n.32, 113 n.36, 148 n.2
Convention of the CIO, 19–22
Converting operations, 64
Corruption charges: answered by Burke, 76, 79–80; answered by Tonelli, 111, 154–55; and "the Dowry," 71–72, 110; and Local 679, 53–54, 81; by Melton, 67; and RFMDA Handbook, 100
Crosby, L. A., 105
Crown-Willamette Paper Co., 25, 28–30, 34 n.40
Crown-Zellerbach Company, 29–30, 34, 54, 140

Dare, John, 31
Davidson, Robert J., 20
Davis, Charles E., 27
Decentralization of AWPPW, 149, 155–56
Delaney, Fred, 123, 133, 140
Democratic Unionism, Program for Militant, 74–75, 80–81, 96
Democracy in unions, xi–xiv, 145; and AWPPW, 148–49; and RFMDA, 96, 115
Department of Research and Education, 43–49, 58, 145
Depression, 8, 11–14, 25, 147
Diamond Match Company, 55
Diana Paper Company, 17
District 50, 22–23
"Dowry, the," 71–72, 110
Dual unionism, xii, 36, 119
Dukes, Clarence W., 45 n.9, 66 n.14, 85 n.45, 88–89

Eagle Lodge, 2
*East Coast Newsletter*, 96, 116
Eastern Conference of Pulp and Paper Mill Unions, 76–77
Electrical Workers, 8 n.57, 9, 120, 155
Employees Association. *See* Pacific Coast Pulp and Paper Mill Employees Association
Escanaba Paper Company, 17
Esposito, Anthony, 20
Ethical Practices Code of AFL-CIO, xi
Everett Local 183, 35, 45, 76, 85 n.45
Everett Local 236, 46
Exploitation of minorities, 59, 71
Eyer, John, 46, 58, 114, 140; defeat of, 113; and regional elections, 51–52; and RFMDA, 107

Farrace, Arthur, 45 n.9, 85 n.45, 86 n.48, 88–89
Federal labor unions, 3

Federal Mediation and Conciliation Service, 140–41
Fibre Board Products, Inc., 25, 34 n.40
Financial responsibility, 97, 104, 155; and the Portland case, 141–43
Finishers Incorporated, 79
Fitzgerald, James F., 3–5
Forfeiture clause in Pulp Workers constitution, 141–43, 149
Fort Edward Labour's Protective Union No. 9259, 3
Fort Edward Local 1, 6, 145
Fort Francis Paper Co., 8 n.55
Foster, J. T., 8 n.58
Fourre, E. P., 27
Frenchtown Local 885, 85 n.45
Funkhauser, Gladys, 26

Gamm, Sara, 117 n
Gear, Carl, 58
General strike in San Francisco in 1934, 31–32
Gilman Local 41, 83, 85
Glenn, Wayne, 84 n.41
Glens Falls Local 20, 85 n.45, 94
Gompers, Samuel, 4
Grasso, Frank, 20–21
Graustein, Jacob, 15
Grays Harbor County Central Labor Union, 26
Grays Harbor Pulp and Paper Co., 34 n.40
Great Northern Paper Company, 15, 32
Green, Stan, 46
Green, William, 25–26
Grievances: and Joint Relations Board, 34; revision in procedures for, 151; role of steward in, 64; and West Coast split, 121, 124
Grimes, Sidney, 61, 123, 125, 134–35
Ground rules for negotiations, 127–34, 148, 150, 153

Hagen, Eugene A., 106
*Handbook for Constitutional Change*, 100–102
Hanna Paper Corp., 8 n.55
Hansen, H. L., 73, 87, 113–14
Hawley Pulp and Paper Co., 34 n.40
Hayes, Paul, 68, 71; and AWPPW's legal actions, 142; dismissal of, 73–74, 76–77; and Portland case, 142; possible reinstatement of, 96, 99, 102, 109–12; and *Truth*, 91
Haywood, Allan S., 20
Holyoke Eagle Lodge, 2
Hetherington, Robert, 57
Hoberg Paper and Fiber Company, 17
Hoquiam Local 169, 26; and NLRB, 40, 146, 152–53; and regional elections, 52; and RFMDA, 87, 95; unrest at, 34–36, 39–41, 45, 48
Howell, Raymond L., 85 n.45
Huggins, Arthur, 27
Hughes, C. L., 109
Humble, Homer, 44
Hunter, Moe, 133–34

IBPM. *See* International Brotherhood of Paper Makers
IBPSPMW. *See* International Brotherhood of Pulp, Sulphite and Paper Mill Workers
Independent unionism, 117
Individual liberties, xiii
Injunctions, use of, 7
Interlake Paper Company, 18
International Association of Machinists, 8–9
International Brotherhood of Electrical Workers, 8 n.57, 9, 120, 155
International Brotherhood of Paper Makers (IBPM), xv; and depression, 10–11; founding of, 2–3; and Pulp, Sulphite Union, 4–5; and training programs, 43. *See also* Internationals; United Papermakers and Paperworkers
International Brotherhood of Paper Makers, Pulp, Sulphite and Paper Mill Workers, 4
International Brotherhood of Pulp, Sulphite and Paper Mill Workers, xiv–xv; and Bureau of Labor-Management Reports investigation, 110–12; and Carlisle group, 6–9; and CIO, 21–22; constitution of, *see* Constitutional change; and depression, 8, 13–14, 147; founding of, 3–4; impropriety in, *see* Corruption charges; and International Paper Company, 18–19, 55; membership of, 14–15, 21; organization activities of, 9–11, 16–18, 25–41, 54; research and education department of, 43–49, 58, 145
International Brotherhood of Stationary Firemen and Oilers, 8, 9
International Brotherhood of Teamsters, 31
International Longshoreman's Association, 31
International Paper Company: and code

hearings, 15; strikes at, 5–8, 146–47; union organization of, 18–19, 55
International Paper Machine Tenders Union (IPMTU), 2
International Research and Education Departments, 43–49, 58, 145
Internationals: challenge to, 27 n.13, 71–97, 120, 126–27, 128, 152; and defeat of RFMDA, 108, 114; and education departments, 47–49; and Employees Association, 37–39, 46; in negotiations, 35, 38–39, 61–65; power of, 38–39, 41, 151, 153; at prewage conference of 1964, 132–33, 148–49; and unfair labor practice charge, 63–66
International Woodworkers of America (IWA), 22, 40, 75, 146
Irvine, Frank, 8–9
Isaacson, Ivor, 55, 89, 122, 130; and Brooks resignation, 81–82; and prewage conference of 1964, 132–35, 148, 153; and education departments, 44, 47–49; and ULA negotiations, 49–51, 56

Job analysis, 46–47, 57
Johnson, George, 3
Joint Relations Board, 34
Jones, Col. C. H. L., 8 n.58
Jones, Judge Robert, 141–43
Jordan, Bob, 85
*Journal* (of the Pulp Workers), 87, 93

Kaehn, Carl, 105
Kapuskasing Local 89, 85 n.45, 86, 94
Kempton, Murray, 71–72, 79–80
Kennedy, Robert, 102
Kerr, Clark, 41
Kimberly-Clark, 22, 93
King, Harold, 85

LaBelle, Maurice, 18
Lake States, 16–18
Lambertson, George, 78, 91, 93, 115
Landrum-Griffin Act, 110, 155
Leavitt, Ralph, 72, 83, 91
Legal actions, 136, 139, 141–144
Leon, Raymond, 110
Lester, Richard A., 145–46
Lewis, John L., 23
Lindsay, Amos, 84 n.41
Lines, Elmer, 132–33, 140
Local 1, Fort Edward, 6, 145
Local 4, Palmer, 145
Local 20, Glens Falls, 85 n.45, 94
Local 23, Berlin, 4
Local 41, Gilman, 83, 85
Local 52, Port Edwards, 105
Local 61, 83
Local 68, Oregon City. *See* Oregon City Local 68
Local 76, Powell River, 85 n.45, 113
Local 89, Kapuskasing, 85 n.45, 86, 94
Local 100, Camas, 106
Local 153, Longview. *See* Longview Local 153
Local 155, 27
Local 161, 27
Local 168, 77
Local 169, Hoquiam. *See* Hoquiam Local 169
Local 183, Everett, 35, 45, 76, 85 n.45
Local 194, Bellingham, 66, 101, 123
Local 230, Capital, 27
Local 236, Everett, 46
Local 242, Portland, 75, 85 n.45
Local 249, Antioch, 45, 48, 50, 85 n.45, 89
Local 309, Bellingham, 36
Local 312, Ocean Falls, 74, 113
Local 318, New York City, 80
Local 375, Philadelphia, 83, 85 n.45, 96, 109
Local 400, Brunswick, 96, 108
Local 433, Vancouver, 45 n.9, 68, 81, 85 n.45, 102
Local 435, Savannah, 105
Local 482, Neenah, 105
Local 580, Longview, 50
Local 679, New York City, 53–54, 58–59, 71, 81, 111
Local 708, Prince Rupert, 52, 85, 87, 102
Local 713, San Joaquin, 45 50, 52, 88
Local 738, Wilmington, 96
Local 850, Mount Diabalo, 55
Local 885, Frenchtown, 85 n.45
Local 947, Arkadelphia, 109
Local power: and autonomy, 65, 148–49, 155; and defiance of Internationals, 148, 152; loss of, 41, 129, 146–47
Longshore strike, 31, 32 n.33
Longview case, 143
Longview Fiber Co., 34 n.40
Longview Local 153, 26; and dissatisfaction, 35–37, 48, 50; suit against, 143; and unfair labor practice charge, 65–66
Longview Local 580, 50
Longview meetings, 27–28
Loomis, Maxwell, 36, 38
Lorrain, Louis H., 44, 72, 74, 78–79;

and Brooks's resignation, 81: and Detroit convention, 103, 107
McCarthy, Senator Joseph, xiii
McCormick, R. B., 74–75, 95, 113
McDonald, David, xiii
McClellan Committee, 53, 71, 111–12, 155
McNiff, John J., 58–59, 92–93, 112
Macphee, Angus, 94, 116; and Detroit convention, 102; and Montreal convention, 67–68; and reform, 76, 85
Malin, John H., 3–5
Management rights clause, 139, 141
Manufacturers Association. *See* Pacific Coast Association of Pulp and Paper Manufacturers
Manufacturers, role of, 150–52
Marathon Paper Company, 17, 57
Marshall, Peter, 87, 116
Meany, George, 138
Melton, Melvin, 84, 85, 132; at Detroit convention, 103; and pensions, 123; and reform, 87, 115; and regional elections, 66–67, 101
Memorandum, anonymous, 117–20, 121, 126
Mercer, Graham, 45 n.9, 85, 87–88, 95, 132, 140
Merger of Paper Makers and Paperworkers, 122, 146
Merger with International Woodworkers, 75
Mersey Paper Company, 14
Michels, Roberto, 118
Militant Democratic Unionism, Program for, 74–75, 80–81, 96
Minnesota and Ontario Power and Paper Co., 8 n.55
Minorities, exploitation of, 59, 71
Monat, C. M., 85 n.45
Mosinee Paper Mills Company, 17
Mount Diabalo Local 850, 55
Munsey, Glenn, 85
Murray, E. B., 8 n.58
Murray, Philip, 20

National Industrial Recovery Act, 14–15, 32–33, 34
National Labor Relations Board, 38; appeal to, 136–38; and Hoquiam, 40, 146, 152; and power of Internationals, 40–41; recent action of, 144; and unfair labor practice charge, 63, 66
National Paper Products Company, 25, 34 n.40
Neenah Local 482, 105
Negro minorities, 59, 71
Newsprint Institute of Canada, 11
New York City Local 318, 80
New York City Local 679, 53–54, 58–59, 71, 81, 111
*New York Post*, 54, 71, 77, 79
Nicol, James P., 43
Northwest Committee for Union Justice, 85, 96

Ocean Falls Local 312, 74, 113
Oliver, Lloyd, 73, 82, 95
Olympic Forest Products Co., 34 n.40
Open Shop, 6–7
Oregon City Local 68, 26, 50, 85 n.45, 135; and Detroit convention, 103, 107; suit against, 143
Oregon Pulp and Paper Co., 34 n.40
Organization: in Canada, 9–11; in the Lake States, 16–18; in the South, 16; on the West Coast, 25–41, 54
Osborne, Ben T., 26–27
Ostling, Wilf, 107
Oxford Paper Company, 22

Pacific Coast Association of Pulp and Paper Manufacturers: and negotiations, 123, 124–28, 134; and pensions, 58, 61–64; and split of AWPPW, 139; and unfair labor practice charge, 58, 63–66
Pacific Coast Council of Paper Makers, 36
Pacific Coast Council of Pulp and Paper Mill Workers, 84–85
Pacific Coast Pulp and Paper Mill Employees Association, 27, 33–36; and ground rules for negotiations, 131–35; and Internationals, 37–39, 118–19; and local autonomy, 118–19, 121; newspaper of, 119; and training programs, 43–46; and Uniform Labor Agreement, 151; weakened power of, 146
Pacific Northwest Conference of Pulp and Paper Employees, 27. *See also* Pacific Coast Pulp and Paper Mill Employees Association
Pacific Northwest Pulp and Paper Mill Employees Association. *See* Pacific Coast Pulp and Paper Mill Employees Association
Pacific Strawboard and Paper Co., 34 n.40
Palmer Local 4, 145
Paper Makers. *See* International Brother-

hood of Paper Makers; United Papermakers and Paperworkers
Paper, Novelty, and Toy Workers International Union, 19, 22
Paper Supply Company, 17
*Paper Worker*, 57
Paper Workers Organizing Committee (PWOC), 20. *See also* United Paperworkers Association
Parker, Oren, 46, 73, 113, 117, 122, 133
Peavy Paper Company, 17
Pension plans: and negotiations of 1959, 61–66, 150–51; and negotiations of 1963, 119, 121, 123–24
Per capita tax, 108, 119
Perrin, William, 50 n.27, 85; and AWPPW, 137, 140, 148; and ground rules for negotiations, 132–33, 135; and proportional representation, 53; and regional election debate, 67, 105; and RFMDA, 86, 95, 103, 115; suit against, 141
Pettebone-Cataract Paper Co., 8 n.55
Philadelphia Local 375, 83, 85 n.45, 96, 109
Phillipps, Shelby, 84 n.41
Phillips, Paul, 65, 120
Port Angeles Local, 45
Port Edwards Local 52, 105
Portland Case, 141–43
Portland Conference, 29–30, 33–36
Portland Local 242, 75, 85 n.45
Powell River Company, 12
Powell River Local 76, 85 n.45, 113
Power: centralization of, 41, 118; fragmentation of, 118, 120, 155–56; of Internationals, 38–39, 41, 66, 129, 151–53; of locals, 65–66, 127, 129, 146–49; of the vice-presidents, 52
Prewage conferences, 38–39, 41, 82, 122, 132–37
Price, Cash, 56, 89, 108
Prince Rupert Local 708, 52, 85, 87, 102
Printing Pressmen, 21–22
Program for Militant Democratic Unionism, 74–75, 80–81, 96
Proportional representation at conventions, 53
Prouty, Senator, Winston L., 110
Publicity at negotiations, 123, 125–28, 130, 131
Puerto Rican minorities, 59, 71
Pulp and Paper Employees Association. *See* Pacific Coast Pulp and Paper Mill Employees Association
Pulp and Paper Workers of Canada, 116
Rainier Pulp and Paper Co., 34 n.40
Randall, Murray, 85
Randall, Roger, 41
Rank and File Movement for Democratic Action (RFMDA): accusation of communism in, 94–95; defeat of, 104, 107, 109, 148–49; formation of, 85–97; and handbook for constitutional change, 100–102; platform of, 86, 95, 96–97, 99, 114–15
Rayonier, Inc., 39
Recall provisions, 96, 99, 107
Recognition clause, 34
Reform movement: early, 35–36, 46; leaders of, 115; program of, 66–67, 75, 95, 96–97, 99, 114–15; support for, 77, 84. *See also* Rank and File Movement for Democratic Action
Regional autonomy, 148, 151
Regional elections: and AWPPW, 149–50; and Detroit convention, 105–107, 114; early sentiment for, 47, 51–52, 66–67, 75; and RFMDA, 86, 96, 100–101
Rehearing clause, 9
Research and education departments, 43–49, 58, 145
Retirement of board members, 67, 75, 96, 102, 107
Retirement plans. *See* Pension plans
Retroactive pay, 140
RFMDA. *See* Rank and File Movement for Democratic Action
Richards, Ray, 17, 58
Riordan Company, 18
Robertson, Oscar, 122–26, 133, 135
Rogers, Henry, 85 n.45; and McNiff expenses, 93; and RFMDA, 92, 97, 115; and *Truth*, 90
Rosebush, George, 85 n.45, 86 n.48, 94
Ruddick, Durwood, 84 n.41
Ruddick, Godfrey: and Brooks's resignation, 81, 84; defeat of, 51 n.32, 113 n.36, 148 n; reelection of, 113, 148; and regional elections, 107; and RFMDA, 92

St. Helens Pulp and Paper Co., 34 n.40
St. Maurice Paper Co., Ltd., 8 n.55
St. Regis Paper Co., 6, 8 n.55, 10, 78, 110
Salem Trades and Labor Council, 27
San Francisco general strike, 31–32
San Joaquin Local 713, 45, 50, 52, 88
Savage, Leo, 85, 86 n.48
Savannah Local 435, 105

Sayre, Harry D., 20
Scarselleta, Mario, Jr., 85 n.45, 86 n.48, 94, 115
Schneider, George J., 26–27, 28–29
Scott Paper Company, 22
Seattle longshore strike, 31
Secret ballot, 103
Segal, Henry: and anonymous memorandum, 117; and Brooks's resignation, 82; at Detroit convention, 104; election of, 68–69, 77; and *Truth*, 91
Senate Select Committee on Improper Activities in the Labor or Management Field, 58
Seniority, 121, 126
Shelton Local, 35
Sherman, John, 33 n.38, 51; and CIO unions, 21; and exploitation of minorities, 59; and Longview meetings, 27; and Portland conference, 34; retirement of, 122; and role of Internationals in negotiations, 35, 39, 49–50; and training programs, 44–47
Sims, W. A., 84 n.41
Skibba, Ruffin J., 105
Smalley, Alex, 78–79
Smethurst, Walter, 20
Somes, Herb, 77
South, the, 16, 92
*Southeast Newsletter*, 96
Southern Kraft Corp., 16, 18–19, 152
Spanish River Pulp and Paper Co., 8 n.55, 10, 26, 28
Steelworkers, xiii, 110, 120, 155–56
Stephan, Jacob, 17
Sterling Paper Company, 17
Sterman, Albert, 90–91
Stewards, 44, 47, 64
Stewart, Charles, 69, 77, 84 n.41
Stone, Chief Justice Harlan F., xi
Strike procedures, 129, 132, 138, 152
Sullivan Herbert, 16, 29–30, 32–34, 51

Thilmany Paper Company, 17
Thunder Bay Paper Company, 10
Tidewater Paper Mills Co., 8 n.55
Tierney, Francis, 32, 91, 110, 117
Tomahawk Paper Company, 17
Tonelli, Joseph, 53; and anonymous memorandum, 117; charges against, 71–74, 77–79, 111, 115, 142, 155; and Detroit convention, 109–12, 115; and "the Dowry," 71–74; and Portland case, 142–43; and RFMDA, 89, 91, 115; and *Truth*, 91
Toy and Novelty Workers, 19, 22
Training programs, 43–49, 58, 145
*Truth*, 90–91

ULA Committee, 136, 141
Unfair labor practice charge, 58, 63–66, 82, 136
Uniform Labor Agreement (ULA), xv, 83, 137; and AWPPW, 140, 141; and ground rules debate, 127–30; and Hoquiam, 39; and International power, 38, 146–47, 151; and legal action, 136, 141; and Manufacturers, 139, 151; and Melton letter, 67; negotiations of, 39, 45, 47, 49–50, 56, 61–65, 121, 124–27; origin of, 33–34; and pensions, 61–64, 124; and proportional representation, 53; and reform, 116, 119, 121; rejection of, 125–26
Union Bag and Paper Co., 8 n.55, 13
Union democracy. *See* Democracy in unions
Union funds, 10, 93, 104, 141–43, 155
Unionism: acceptance of, 31–33; dual, xii, 36, 119; independent, 117; Program for Militant Democratic, 74–75, 80–81, 96; start of, 2
Union membership, 14–15, 21
Union security clause, 140
Union shop, 140–41
United Auto Workers (UAW), 120, 156
United Brotherhood of Carpenters and Joiners of America, 8–9
United Brotherhood of Papermakers, 2
United Gilman Local 41, 85
United Mine Workers (UMW), 22–23
United Papermakers and Paperworkers, xiv, 3 n.14, 14, 58, 61–65, 122
United Paper, Novelty and Toymakers International Union, 19, 22
United Paperworkers of America (UPA), 20–22
United Steelworkers, xiii, 110, 120, 155, 156

Vancouver Local 433, 45 n.9, 68, 81, 85 n.45, 102
Van Deusen, Floyd, 108
Victor, Reynold, 49
*Voice of the Program*, 81

Walker, William, 106
Washington Pulp and Paper Corp., 34 n.40
Washington State Federation of Labor, 25
Wells, Burt, 77; and Brooks's resignation, 83; and McNiff, 93; and reform,

75, 80–81, 115, 121; and RFMDA, 85–86, 116
Wentz, Keith, 113
West Coast newspaper, 119
West Coast organization, 25–41, 54
Western Association. *See* Association of Western Pulp and Paper Workers
West Virginia Pulp and Paper Company, 5, 20, 22
Weyerhauser Timber Company, 25–26, 28, 34 n.40
Williams, Wilbur, 107
Wilmington Local 738, 96
Wirtz, Willard, 140
Wolf, Robert B., 26, 28–29
Woodworkers. *See* International Woodworkers of America

Yellow-dog contract, 5
Young, C. O., 26

Zellerbach, J. D., 29–30, 34